W9-BOL-520

Photographic Guide to the Butterflies of Britain and Europe

Photographic Guide to the Butterflies of Britain and Europe

Tom W. Tolman

OXFORD
UNIVERSITY PRESS

OXFORD
UNIVERSITY PRESS

Great Clarendon Street, Oxford OX2 6DP

Oxford University Press is a department of the University of Oxford.
It furthers the University's objective of excellence in research, scholarship,
and education by publishing worldwide in

Oxford New York

Athens Auckland Bangkok Bogotá Buenos Aires Calcutta
Cape Town Chennai Dar es Salaam Delhi Florence Hong Kong Istanbul
Karachi Kuala Lumpur Madrid Melbourne Mexico City Mumbai
Nairobi Paris São Paulo Shanghai Singapore Taipei Tokyo Toronto Warsaw

with associated companies in Berlin Ibadan

Oxford is a registered trade mark of Oxford University Press
in the UK and in certain other countries

Published in the United States
by Oxford University Press Inc., New York

© Tom Tolman 2001

The moral rights of the author have been asserted
Database right Oxford University Press (maker)

First published 2001

All rights reserved. No part of this publication may be reproduced,
stored in a retrieval system, or transmitted, in any form or by any means,
without the prior permission in writing of Oxford University Press,
or as expressly permitted by law, or under terms agreed with the appropriate
reprographics rights organization. Enquiries concerning reproduction
outside the scope of the above should be sent to the Rights Department,
Oxford University Press, at the address above

You must not circulate this book in any other binding or cover
and you must impose this same condition on any acquirer

A catalogue record for this book is available from the British Library

Library of Congress Cataloging in Publication Data

Tolman, Tom.
Photographic guide to the butterflies of Britain and Europe/Tom W. Tolman.
Includes bibliographical references (p.).
1. Butterflies–Great Britain–Identification 2. Butterflies–Europe–Identification. I. Title
QL555.G7T65 2001   595.78′9′094022–dc21   00-049203

ISBN 0 19 850607 4 (Hbk)
ISBN 0 19 850606 6 (Pbk)

Typeset by Florence Production Ltd, Stoodleigh, Devon
Printed in Italy on acid-free paper by Vincenzo Bona s.r.l., Turin

Preface

In his early days of scientific training, the author was heavily imbued with the importance of establishing, at the outset, the clear, primary purpose of a project. He was told . . . in so many words . . . state your objectives clearly and identify all possible pathways leading to their achievement.

In regard to its objective, this modest volume may, to its user, present itself as a means of finding and identifying butterflies. Whilst this work has, of course, been prepared with that as one of its purposes, which, in more basic terms, may be defined as the pursuit of pleasure, it is not its primary purpose. Life's pleasures arise in many and various ways, one of which derives directly from our children and grand-children. Indeed, it is their interests and needs which generally concern us most. In discharging the responsibility of ensuring that the quality of their lives remains at least equal to ours, we are obliged to pass on our privileges to our heirs – the enjoyment of looking at butterflies being but a single example. Providing this particular pleasure, however, serves a multitude of other purposes. Just as we need a home, butterflies need a place to live; such areas are most often referred to as habitats. These are invariably shared with other living things and, indeed, the vast majority of all land-based animals and plants pervading our planet live within the domain of the world population of butterfly species. Damaging a butterfly habitat will, in consequence, almost inevitably, affect the homes of a vast array of organisms other than butterflies. Of course, it is possible that our children will not share our enthusiasm for butterflies, but, instead, develop an interest in birds, beetles, snakes, or spiders, or, indeed, no living organism other than their own kind. Whatever their preference, it falls to us to perpetuate the choice we would wish for ourselves. In deference to our large

commitment to the care of children, can we, in all conscience, look into the innocent and trusting eyes of our progeny and tell them that we have, through selfishness or carelessness, eliminated, from their lives, a potential source of great pleasure? The answer to this question proclaims the primary purpose of this work – to encourage the conservation of every element of our natural heritage for the benefit of those we hold most dear, if not for the creatures we threaten. From this flow all other objectives, and, indeed, the very method by which all associated aims may be achieved. As the author is not the inventor of the method, he is permitted to commend it for its elegant simplicity and obvious efficacy: most simply stated, looking after wildlife habitats requires, in principle at least, doing nothing more than nothing. Butterflies will survive if their homes are left alone – they did, after all, manage well enough for millions of years before the questionable quality of our behaviour created the concept of conservation. If their collective habitats, and, by intimate association, those of all other species, including our own, are to be secured, then we are obliged to alter, and alter very quickly, our approach to personal, present-day problems. When we tell a young child who is about to descend some steps to 'be careful – they might be slippery', we are advising caution in the face of the unknown. Nobody can, with reliability, predict the ultimate outcome of the many experiments we have, for many years, been making with the lands, seas and atmosphere of our planet. To avoid the potentially dire consequences of continued indiscretion, perhaps our best interests are best served by taking the same good advice we thrust upon our children.

In the complete and final analysis, the battle for butterflies or, more particularly, butterfly habitats, is a battle we cannot afford to lose.

TWT

Acknowledgements

Inevitably, no work of the present kind is possible without the collective efforts of many researchers. As the scientific literature provides, accordingly, the basis for this volume, the author is greatly indebted to those responsible for generating this indispensable source of data. For helpful discussion and providing updated distributional data, the author has pleasure in extending his thanks to: Marco Chiavetta, John Coutsis, Roberto Crnjar, Roland Essayan, Derek Gibson, Zdravko Kolev, Lazarus Pamperis and Rudi Verovnik.

For their considerable generosity in providing supplementary photographs, the author is deeply indebted to the following photographers: Stoyan Beshkov (*D. chrysippus*, *E. cassioides*); Martin Gascoigne-Pees (*K. psylorita*, *H. cretica*); David Jutzeler (*P. phoebus*, *L. helle* (female upperside), *A. artaxerxes allous*, *A. thersites* (male upperside), *A. iris*, *N. antiopa*, *B. pales*, *C. thore* (female upperside), *M. diamina* (male underside), *E. intermedia*, *M. arge*, *S. ferula* (male upperside), *E. manto*, *E. claudina*, *E. flavofasciata*, *E. christi*, *E. pharte*, *E. tyndarus*, *E. nivalis*, *E. ottomana* (S France), *E. scipio*, *E. styx*, *E. neorides* (male upperside), *P. tithonus*, *C. darwiniana*, *P. aegeria* (female upperside), *L. achine*); Josquin Lafranchis (*C. minimus* (female upperside), *C. argiolus* (male upperside), *P. aegeria* (female underside)); Tristan Lafranchis (*P. brassicae* (SW France), *E. alcetas* (female upperside), *I. iolas* (male upperside), *P. argus* (male upperside/female underside), *A. agestis* (female upperside), *A. dolus*, *A. ripartii* (male upperside), *I. io*, *A. urticae*, *M. diamina* (male upperside), *E. neorides* (female underside), *P. aegeria* (male upperside), *L. achine*); Mike Majerus (*C. nastes*, *H. iduna*) and Salvatore Spano (*P. hospiton*).

Finally, for her willing and tireless assistance in all aspects of field-study, photography and the preparation of text, it is with particular pleasure that the author expresses his deep and heartfelt gratitude to his wife, Sally.

Contents

Abbreviations

f.	form
fw	fore-wing
fwl	fore-wing length (wing-apex to wing-base)
gc	ground colour
hw	hind-wing
LHP(s)	larval host plant(s)
Mt/Mts	Mount or Mountain/Mountains
Mte	Monte (Mountain)
pd	post-discal
s	space (an area of wing-membrane between the veins)
sp./spp.	species (singular) / species (plural)
ssp./sspp.	subspecies (singular) / subspecies (plural)
unf	fore-wing underside
unh	hind-wing underside
uns	fore-wing and hind-wing undersides
upf	fore-wing upperside
uph	hind-wing upperside
ups	fore-wing and hind-wing upper sides
v	vein
var.	variety

Introduction

Arrangement of main text

All European butterflies belong to one of the following families arranged in the following order:

Papilionidae
  Swallowtails, Apollos, Festoons
Pieridae
  Whites, Sulphurs, etc.
Lycaenidae
  Blues, Hairstreaks, Coppers
Riodinidae
  Metalmarks
Libytheidae
  Snouts
Danaidae
  Monarchs, Milkweeds
Nymphalidae
  Admirals, Gliders, Fritillaries, etc.
Satyridae
  Browns, Graylings, Ringlets, Heaths
Hesperiidae
  Skippers

The account for each butterfly species starts with its scientific (Latin) name, which comprise genus and species names in accordance with the binomial system devised by the great Swedish biologist, Carolus Linnaeus. English names are also given where, as is usually the case, one is available. A genus is a group of closely related species, whilst a species is a group of individuals able to interbreed freely to produce healthy, viable offspring. The scientific names used here are those used by the present author in the *Collins field guide to the butterflies of Britain and Europe*. Although the generic classification adopted will be found to differ from other publications, this need cause little if any confusion since species names almost always remain unchanged. Indeed, the main reason they are changed is dictated by a requirement that genus and species names belong to the same gender, that is, masculine, feminine, or neuter. The merit of this dictate seems highly questionable. If it does become necessary to place a species in a different genus, there seems no compelling reason to risk confusion by altering its name in blind, pedantic attachment to a technical requirement of the Latin language. That genus and species are not only scientific names, but, in effect, scientific definitions argues strongly for the priority of scientific need. Far from being a tool for the creation of confusion, language exists solely to serve the need to communicate – with clarity whenever possible.

Distribution

The geographical area covered is shown in Figure 1, and comprises the geopolitical region of Europe (including the Canary Islands, the Azores and Madeira) except the countries of Belarus, Moldova and Ukraine. Iceland has no indigenous butterflies. The map adjacent to each species account in the main text, is intended as a rough, graphical summary of distribution within the region covered. It should be noted that a species will generally not be found at all localities within the boundaries given (shown in a green tint). For migrant species, the average extension of range resulting from migration is indicated (in a pink tint). The extent of migration, from year to year, may vary considerably, mostly according to seasonal weather conditions. The altitudinal range, in company with other information, is often of great value in locating a species. It is also sometimes useful in aiding the identification of similar species occurring in the same geographical region but separated by altitude. Altitudinal range often accounts for recurrent distributional patterns shown in the maps: whilst some butterflies may be restricted to higher, and therefore cooler regions in one or more mountain ranges, others may be confined to hot, low-lying coastal areas of the Mediterranean. Another interesting pattern of distribution relates to species occurring at high altitudes in, for example, the Alps or Pyrenees, and at or near sea-level in the Arctic region. Owing to the relationship between temperature, altitude and latitude, such geographically well-separated populations live under closely similar climatic conditions.

Figure 1

Description and variation

Description

As photographs provide the main means of identification, descriptions of wing-markings are generally limited to wing-surfaces not shown. However, clearly visible diagnostic features are given in all cases, and particular attention is drawn to those which may be helpful in distinguishing closely similar species. Distinctive external characters other than those relating to the wings are given where apparent for a butterfly in its resting pose. The major, clearly visible external features of a butterfly's anatomy are shown in Figure 2. Generally, the only easily discernible difference between males and females, apart from wing-markings where different, is the

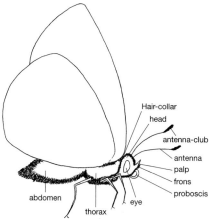

Figure 2

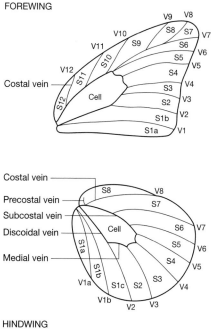

Figure 3

shape and size of the abdomen, which in the female, as it contains the eggs, is usually noticeably 'plumper'.

To aid the formalized description of wing-markings, each wing-surface is most conveniently divided into the specific areas shown in Figure 3. The location and notation of wing-veins and intervening spaces is shown in Figure 4. The combination of these standard descriptive systems is entirely logical and allows all parts of all wing-surfaces to be described without confusion.

In the males of many species, specialized scales known as androconia (also referred to as androconial scales or scent-scales) are sometimes grouped into patches (in various shapes and sizes), tufts or conspicuously concentrated along veins. Androconial patches, often called sex-brands, may be useful in determining sex in the absence of other distinctive features. An androconial scale itself possesses special cells containing chemicals called pheromones which are released by the male in courtship to attract the attention of the female.

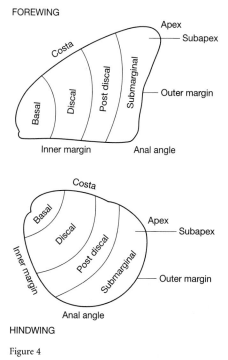

Figure 4

Measurements of butterfly size, often and most conveniently expressed as fore-wing length, have been omitted. As an aid to identifying butterflies in the field, the impression of size overrides any thought of wing length simply because size is judged, instinctively, by wing-area – not wing length. In practice, it will be found that very little experience is required to gauge, at a glance, the relative sizes of different species, groups, or families of butterflies.

Variation

This deals mainly with differences in wing-markings and size arising from the different average conditions under which a butterfly lives in different parts of its geographical or altitudinal range. Where known or suspected, the causes of such variation are indicated. This information can be of some use in identifying a butterfly which may not entirely match the description of its normal form. The differences which often arise between generations (broods), where two or more are produced in the same year, appear to relate largely, if not entirely, to the effect of changing seasonal conditions of temperature, humidity and day-length (photoperiod) on larval and/or pupal development. The influence of temperature or moisture or both is also seen as a variation between years due to abnormal weather conditions: such effects are perhaps most apparent following a late cold spring.

Although moderately higher temperatures tend to accelerate larval development, thereby producing larger butterflies, very hot conditions sometimes have the reverse effect where desiccation of host plants reduces the nutritional quality of the food source. However, this is not always so, and sometimes a species living at high altitude is much larger than might be expected. It seems probable that the lower temperatures, prevalent at high altitude, retards early-stage development to the point where the butterfly cannot complete its life-cycle in a single year and is forced to hibernate part-way through its larval stage. Although development continues slowly after hibernation, the extra time available for the completion of growth results in a bigger larva.

Recurrent colour variants in the female of some species, notably the Clouded Yellows, appear to be genetic in origin.

All forms of variation noted above may be described as forms. The term is arguably the most useful and generally the most appropriate formal description of variation below species level. Nevertheless, the term subspecies is widely used, especially for well-separated populations for which a marked and sensibly constant difference is apparent.

Excluded from the types of variation discussed above is that which occurs between individuals within a single population. However similar they may be, no two individuals of the same species will be identical. This simple fact has far-reaching consequences. If the range of a species is large, a strong possibility exists that it will occur in different kinds of habitat or be subjected to different regimes of climatic conditions – temperature, rainfall, levels of sunshine, and so on. As the reader will doubtless know, all organisms adapt to the conditions under which they live, and, in direct consequence, produce local or regional races, the best example being that of our own species. If the conditions under which an organism lives in two different environments diverge progressively, perhaps due to climatic change over time, a corresponding change in the character of the two populations may be sufficient to create a new species. This progressive adaptation equates, of course, to the process of evolution. It is made possible by the genetic variation between individuals in any one population. The mechanism is easily understood. Those individuals carrying the genes which provide the characters best suited to survival are the ones that survive – along with their genes – to produce a new generation, whilst those less suited to survival are progressively eliminated from the population, again along with their genes. As the group of individuals comprising the new species retain the same or similar level of variation in their pool of genes, albeit a different gene pool, further adaptation is possible, if and when it becomes necessary.

Flight-period

The overall period or periods for which a butterfly is expected to be on the wing is given. However, flight-period may vary widely within the area of distribution. As a rule, species occurring at low altitudes in more southerly regions will emerge earlier than those in their northern range or at higher altitude.

Emergence dates – the beginning of the flight-period – are also often influenced by seasonal weather conditions. A late cold spring in Arctic regions, for example, can delay the emergence of some species by more than a month. For the purpose of locating a particular sex of a butterfly, it is helpful to know that the females of most species emerge later than the males, in some cases by more than a week. The number of broods per year is also variable in some species. Whilst a single brood may be produced in more northerly locations or at higher altitudes, the same species may produce two or more broods in warmer regions or at lower altitudes.

Habitat

The main features of the butterfly's habitat are described with the aim of helping users locate butterflies. Where applicable, attention is drawn to the variety of habitat types in which a species may occur. Knowing about habitat can often be enough to find a species even when it is not flying, that is, when it is at rest in dull weather or settling down to roost in late evening. Knowledge of larval host plants is especially useful and, along with a knowledge of habitat, it is even possible to discover new colonies of butterflies by locating their early stages (ova, larvae, or pupae). Both English and scientific names are given for host plants.

Behaviour

Typical behaviours which may aid identification are described. Some species, especially the 'blues' and 'skippers', often gather in large numbers on damp ground by springs or streams, where they may remain motionless for long periods. They may then be examined at leisure, at close quarters, making identification much easier. The way a butterfly flies is also a useful identification guide since it is usually typical of the family or genus to which it belongs: for example, the slow, up-and-down, 'jerky' flight of a 'brown' distinguishes it immediately from the relatively fast and more direct flight of a hairstreak of similar size and colour. Very low and often extremely rapid – 'buzzing' – flight is typical of a skipper. As females of many species spend much of their time crawling amongst their

host plants in search of egg-laying sites, they are often much less in evidence than the males: in such cases, a knowledge of larval host plants serves the dual purpose of locating females as well as the habitat itself. The kinds of plants, including trees, to which a butterfly is attracted is often a further useful indicator of family, and sometimes even the species. For finding certain butterflies, it may be helpful to know how their behaviour changes during the day. A species may, for example, bask in early morning sun, feed on flowers, or enter into courtship until the hottest part of the day, at which time it may seek the shade of leaves in the upper branches of a tree, thereby becoming less easily detectable. For some species, males may sometimes be found on mountain summits – quite literally at the highest point – and far removed from their breeding grounds. The reason for this curious behaviour, known as hilltopping, is by no means clear. An interesting behaviour of some well-known butterflies, for example the Milkweed, Painted Lady, Clouded Yellow, and Red Admiral, is that of migration. This relates, so it seems, to a survival strategy based on the principle of not putting all eggs in one basket. Unlike a species which lives in a small community from which it rarely strays, migrants will, with seasonal regularity, extend their range in spring and summer, often very considerably, by dispersing into new areas and establishing new breeding colonies from which further migration can occur. The extent to which this happens depends very much on weather conditions. With the onset of winter, most, if not all, of the newly established colonies die out with the loss of perhaps millions of insects. Nevertheless, the species itself survives, and in the event of unfavourable climatic change in its permanent, present-day habitats – inevitable in the very long-term – it is almost certain to be able to continue its existence somewhere.

Conservation

Information given here relates only to butterfly species which are considered to be under a particular threat. Most butterflies, especially common and wide-ranging species, appear to be relatively secure, at least for the present. However, through continued interference with, or destruction of, the natural environment, this situation is likely to change: indeed, many of today's rare or endangered butterflies were, not so very long ago, relatively common and free from threat. These facts raise an issue of paramount importance in the matter of butterfly conservation. There can be no doubt that for butterflies in general and endangered species in particular, their protection can only be secured by protecting the places in which they live – their habitats. On this crucial point – already made in the Preface – agreement seems quite universal. There is also little doubt of the need for urgency in complying with this imperative.

At the risk of appearing cynical, the author wishes to express the opinion that much legislative protection afforded butterflies is a facile, cost-free political response to pressure from conservation groups. Protecting butterflies whilst leaving open the options on the use of their habitats seems, at best, to reflect ignorance and, at worst, insincerity. One of the more insidious dangers of this circumstance is that of lulling the uninitiated into a false sense of security, thereby relieving the political pressure so urgently required. The solution to this problem is as easy to see as it is to apply – at least in principle. It is based on the premise that nothing motivates like self-interest, and if politicians wish to retain power they are obliged at least to try to do what we, the ordinary citizens, ask of them. Ultimately, the power to protect our planet is in our hands – what we do with it is our choice and ours alone.

Whilst a nature reserve testifies to our failure to provide adequate living space for wildlife, a willingness to act upon concern for a threatened species is a step in the right direction. Where possible, expanding nature reserves by reclaiming adjacent territory and establishing and maintaining corridors of communication to other, similarly extended sites are the logical steps required to enhance further the prospects for butterflies and other inhabitants. One of the disadvantages of reserves is the restriction of access sometimes imposed on the general public. This may well be justified for very small and ecologically fragile sites to which large numbers of people may be attracted and cause inadvertent damage – all the more reason, so it would

seem, for having bigger reserves and more of them. Of course, reducing the reason for their very existence by protecting what remains of the unprotected natural environment is, by far, the preferred course of action.

How to use this book

For the reader with little previous knowledge of butterflies, there can be no doubt that some effort to gain familiarity with the different types of butterflies, their habitats, behaviour, and so on will be well rewarded when faced, for the first time in the field, with the task of identification. The practice of doing some 'homework' may be refined in many and various ways. If, for example, one is going to a particular place for a holiday to look for butterflies, making a list of species known to occur in the region is certainly worthwhile. This may be done, quite easily, by systematically inspecting the distribution maps adjacent to the description given for each species in the main text. The altitudinal range within geographical areas is not only of value in locating a butterfly, it is occasionally very helpful as a means of separating similar species whose altitudinal ranges are not known to overlap. For checking altitude, a good map is an obvious advantage. It should, however, not be assumed that all is known about butterfly distribution. It is always best to expect the unexpected: indeed, it is worth remembering that every butterfly species in every locality was observed by someone for the first time. The very existence of distribution maps is, of course, due directly to the cumulative efforts of a great many individual observers. To improve our knowledge, new observations should always be reported.

To maximize the benefit of preparing for a field-trip, homework should extend to all sources of information given in the main text. Using a systematic approach, a list of all species likely to be encountered is easily prepared. Comparing flight-periods may well be enough to separate two similar species from the same habitat if they do not fly at the same time. A knowledge of habitat is especially valuable in locating butterflies and often provides clues as to the possible identity of a butterfly which might have been poorly observed.

Gaining at least some familiarity with larval host plants is a considerable advantage, especially for butterflies dependent on a single plant species. Apart from providing a check on the suitability of a habitat, eggs are more usually laid on host plants, which means, of course, that they will visited by female butterflies at some stage. Indeed, females of many species will be found only in close proximity to the host plants upon which they lay their eggs. It should be noted that not all females lay their eggs on the plants upon which their larvae subsequently feed: they may instead choose other living plants, dead vegetation, or even stones and rocks. Whilst males generally range more freely, the obvious need for contact between the sexes often occurs close to host plants.

Even a very small patch of damp soil in an otherwise hot and arid landscape is, to a butterfly, an oasis, a place where butterflies and other insects often assemble in large numbers to drink. For the butterfly observer, this is an excellent opportunity to watch and study at very close quarters, for under these circumstances butterflies are mostly very reluctant to fly: indeed, attempts to provoke flight by poking or prodding are not uncommonly quite futile. If they are disturbed, they usually return within seconds, or minutes at the most. Almost without exception, butterflies found on damp ground, usually the muddy or sandy margins of rivers, streams, springs, etc., will be males. It appears that water itself is less important than the dissolved salts which are a peculiar physiological requirement of the male. Both sexes, however, require energy for flight, usually provided by the sugars contained in the nectar of flowering plants. Nectar also provides the water required to replace that lost in normal activity. For the female, other nutrients contained in nectar are acquired for egg development. Seeking out and observing nectar-rich plants, which, for the above reasons, are especially attractive to butterflies, is always worthwhile. Whilst, in this regard, the value of brambles and thistles will probably be known to many, the peculiar potency of the buddleia bush – commonly referred to as the 'butterfly-bush' – surely stands out in the minds of most as the earliest childhood encounter with some of the largest and most colourful of the European butterflies. In southern Europe, where the buddleia bush is rarely seen, one or more of the many species of thyme will be found to have a similar attraction. Moreover, these generally very small, low-growing or prostrate herbs will often be found in hot, arid regions where few other nectar-rich plants can survive. The 'secret' of their success is a robust and deep tap-root that enables the plant to obtain water in the driest of summers. In such circumstances, a small fresh patch of thyme, growing on hard, sun-baked ground or amongst rocks uncomfortably hot to the touch, comprises another form of oasis and one well deserving the attention of the butterfly enthusiast. In the many and varied butterfly habitats in Europe, numerous other nectar-rich plants are, of course, to be found, the comparative value, if not the identity of which, will be as easily determined.

In attempting a field identification, focusing very intently on all on relevant features, wing-markings, size, colour, shape, etc., will greatly improve the chances of recalling such detail once the insect has flown – which, needless to say, it will usually have done long before its picture has been located. The latter event will, incidentally, occur more often if the would-be observer stands in close proximity to the butterfly of interest whilst frantically flicking or fumbling through the pages of a book! Butterflies are programmed to respond immediately to threat, and, as the reader will know already, sudden movement in its field of vision is the surest way of putting the creature to flight. A stealthy, unflustered approach and concentrated attention are the keys to success. In turn, the key to concentration is practice: the process is exactly like the parlour game in which one is required to recount, with eyes closed, all objects seen in a room during a brief period of observation. Here, it is worth noting that whilst the eyes do the 'looking' – only the brain can deal with the immeasurably more complex business of 'seeing'. There are many illustrations of the importance of the need for this distinction, but perhaps the best, and possibly the one most familiar to addicts of the very fine natural history television programmes, relates to that of a perfectly camouflaged moth at rest. When we are shown the bark of a tree in close-up, but without comment, for most of us the bark of a tree is all we see. It is only when asked if we can see the moth that we begin to concentrate, whereupon the vague outline of the insect suddenly appears. Once the brain begins to correctly process the information that the eyes are already and quite unavoidably providing in full, the true picture emerges. Eyes are, of course, truly marvellous devices but they are exclusively optical sensory organs, designed to detect light in an organized fashion. Our brains, on the other hand, are much more versatile and once programmed, can, like computers, perform all manner of wonderful tasks. The process of programming is simplicity itself – it is merely a matter of motivation. Once a keen interest develops, the brain will automatically do what is required of it, and the skill of 'seeing' will improve with practice.

The misfortune of initially failing to identify a butterfly can usually be reversed with a little patience. There is a good chance that the insect will not fly far before re-alighting, and indeed, many butterfly species are in the habit of returning to their original perches after a brief time. Also, of course, other specimens of the species are almost certain to be present in the same locality. One quite indispensable piece of advice is never, under any circumstances, avert one's gaze from a resting butterfly once it has been detected from a distance. Even the experienced are recurrently astonished to find that an insect observed a fraction of a second earlier has since vanished without trace! If, say, it was sat on a leaf in a bush, it was probably still there, but re-locating a particular leaf amidst a myriad of others can be a challenge. The result of approaching and inspecting the bush too closely is as inevitable as it is predictable! The problem may worsen if the butterfly closes its wings and, in so doing, acquires the very appearance of the leaves amongst which it rests. Butterflies, it should be remembered, are designed to evade the attention of predators such as birds and lizards.

Although the above suggestions for homework may seem a little involved, it is probably easier to do than it is to describe. If initial efforts to identify butterflies seem daunting, much comfort is gained by recalling the problem of trying to ride a bicycle for the first time! It might also be noted than even experienced butterfly specialists occasionally have difficulties in butterfly identification – difficulties which, on occasions, have resulted in the discovery of new butterfly species. Unfortunately, it is a fact that some species cannot be identified reliably without resorting to dissection. In regard to the problems of field identification in such instances, the novice and the knowledgeable become quite equal.

Papilionidae

The 11 European representatives of this family comprise some very large, distinctive, and colourful butterflies, of which the Swallowtails are perhaps the most familiar. Sexes are easily distinguished by generally heavier markings and the larger abdomen on the female, which, for the Apollos is further indicated by the presence of a prominent white structure, known as a sphragis, formed near the tip of the abdomen after fertilization to prevent further mating.

Papilio machaon Swallowtail

Distribution • Europe to N Fennoscandia and most Mediterranean islands. Absent from Atlantic Islands & British Isles, except for a very restricted area of SE England (Norfolk). 0–3000 m, usually below 1500 m. Widespread but usually local and uncommon.

Description • Very distinctive – unlikely to be confused with other species.

Flight-period • February–October in 1–3 broods, according to locality.

Habitat • In N & C Europe, damp or dry places. In Mediterranean region, hot, dry places, often areas of cultivation. LHPs include many species of Umbellifer and Rue families (Apiaceae & Rutaceae), especially fennel (*Foeniculum vulgare*). In England, exclusively milk-parsley (*Peucedanum palustre*).

♀ Samos, Greece

Behaviour • Shows strong dispersive/migratory tendency. Flight fast and powerful. Males sometimes visit mountain summits, where they may also assemble in small numbers and remain flying around highest point – quite literally – for some hours.

Conservation • Under no particular threat throughout most of range. Much more vulnerable near limit of range, where colonies may be very small and isolated, e.g., England, where its continued existence is probably due entirely to rigorous protection of habitat as well as insect itself.

Papilio hospiton **Corsican Swallowtail**

Distribution • Known only from Corsica & Sardinia. Generally 500–1200 m, but records range from shore-line to high mountain summits. Very sporadic, generally associated with LHPs.

Description • Resembles *P. machaon* but quite distinctive in overall appearance. Ups & uns black markings more extensive; uph pd blue spots in a regular, well-developed series; 'tail' at v8 much shorter.

Flight-period • Generally mid-May to late July in one prolonged brood. Records span mid-March to mid-August.

Habitat • Grassy hillsides and valleys, often amongst bushes and rocks. LHPs: Corsican rue (*Ruta corsica*) and giant fennel (*Ferula communis*).

Behaviour • Males often visit mountain-tops. Females also wander extensively but only in pursuit of nectar-rich plants and egg-laying sites.

Conservation • Although protected by legislation, seems under no particular threat.

♂ Sardinia

Papilio alexanor **Southern Swallowtail**

Distribution • SE France. NE & SW Italy. NE Sicily. W & SW Balkans. Greece, including Corfu, Kefalonia, Lesbos, and Samos. 0–1700 m. Extremely sporadic and very local.

Description • Uph & unh with black discal band. Readily distinguished from *P. machaon* and *I. Podalirius* in overall appearance.

Flight-period • Mid-April to mid-July in one prolonged brood.

Habitat • Very distinctive. Hot, dry, rocky, often precipitous slopes, usually of limestone. LHPs: several species of Umbellifers, including *Ptychotis saxifraga*, honewort

(*Trinia glauca*), burnet saxifrage (*Pimpinella saxifraga*), giant fennel (*Ferula communis*), and wild parsnip (*Pastinaca sativa*).

Behaviour • Both sexes exploit air currents in aiding flight and often visit flowers of red valerian (*Kentranthus ruber*).

♂ Z N Greece

Iphiclides podalirius **Scarce Swallowtail**

Distribution • Spain to English Channel & Baltic coast, throughout S Europe & most Mediterranean islands including Corsica, but reportedly absent from Sardinia. 0–1500 m. Widespread, locally common.

Description • Overall appearance distinctive – easily separated from related species.

Flight-period • March to early October in 1–3 broods according to locality and altitude.

Habitat • Hot, and often very dry bushy places, woodland margins, and orchards. LHPs: in natural, wild habitat, principally blackthorn (*Prunus spinosa*), hawthorn (*Crataegus monogyna*), mountain ash (*Sorbus aucuparia*); in orchards, most species of cultivated fruit trees, especially of plum family (*Prunus*).

Behaviour • Adults are greatly attracted to nectar-rich shrubs and trees. In hovering or gliding fashion, both sexes appear to exploit air currents in aiding flight.

Conservation • Becoming increasingly scarce in C Europe, due possibly to damage to woodland margins and hedgerows – commonest habitat of LHPs.

♀ NE Greece

♀ Samos, Greece

Zerynthia polyxena **Southern Festoon**

Distribution • SE France, Italy, including Sicily, through S Switzerland. SE Poland to European Turkey & Greece. 0–1700 m, generally below 900 m. Widespread but local.

Description & variation • Upf *without* red spots in s1b, s4–6, s9, or cell (cf. *Z. rumina*). § S Europe. Female ups sometimes ochreous. § S France & Italy. Ups black markings more extensive, upf red spot usually absent in s9.

Flight-period • Late March to early July in one prolonged brood.

Habitat • Hot, dry, grassy and bushy, often rocky places; common in areas of neglected cultivation. LHPs: birthworts, including birthwort (*Aristolochia clematitis*), round-leaved birthwort (*A. rotunda*).

♂ N Greece

Zerynthia rumina **Spanish Festoon**

Distribution • Iberia. S France. 0–1500 m, generally below 1000 m. Sporadic, locally common.

Description & variation • Resembles *Z. polyxena*. All red markings better developed but variable. Female ups gc sometimes ochreous yellow.

Flight-period • Generally late March to May in one prolonged brood, but records span February–October, indicating second brood in S Spain.

Habitat • Hot, dry, rocky places, often amongst scrub; also, cultivated areas and dry, flowery meadows. LHPs: birthworts, including *Aristolochia pistolochia* and round-leaved birthwort (*A. rotunda*).

♂ S Spain

♂ S Spain

Zerynthia cerisy **Eastern Festoon**

Distribution • S Balkans & N Greece. Aegean islands of Lesbos, Chios, Samos, Kos, Simi, Rhodes, Kastellorizo, & Crete. 0–1100 m. Very sporadic & local.

Description & variation • Female gc sometimes ochreous yellow. § Rhodes, both sexes. Red spots usually replaced with orange. § Crete. Smaller; markings reduced; hw outer margin more rounded.

Flight-period • Mid-March to late July in one prolonged brood.

Habitat • Hot, dry, grassy, bushy places; often in river valleys and cultivated areas. LHPs: birthworts, including (*Aristolochia clematitis*).

♀ Samos, Greece

♂ Samos, Greece

♂ Crete, Greece

Archon apollinus **False Apollo**

Description • Very distinctive in overall appearance.

Distribution • Bulgaria. Greece: districts of Thessalonika and Alexandroupolis, Lesbos, Chios, Samos, Kos, Rhodes. European Turkey. 0–1100 m. Very local, generally common.

Flight-period • March–April in one brood.

Habitat • Cultivated areas (orchards, vineyards, etc.). LHPs: birthwort (*Aristolochia clematitis*).

Behaviour • Often rests with open wings at ground level in sunny or warm, overcast conditions.

Conservation • Has suffered extensive local extinction due to chemical spraying in cultivated areas, main residual habitat.

♂ Samos, Greece   ♀ Samos, Greece

Parnassius apollo **Apollo**

Distribution • Most larger mountain ranges from Spain to S Fennoscandia, Balkans, & Greece, including NW Peloponnese. Absent from Mediterranean islands except Sicily. 500–2400 m, generally above 1000 m in S Europe. Sporadic, often locally abundant.

♂ N Greece

♀ N Greece

Description & variation • Resembles *P. phoebus*. Male upf without red pd spot in s8. Antennal shaft pale grey, narrowly ringed darker grey (cf. *P. phoebus*). Markings regionally and locally variable. § S Spain & S Greece (Mt Erimanthos). All red markings replaced by dull yellowy orange.

Flight-period • Early May to September in one brood. Emergence date depends on locality.

Habitat & behaviour • Dry or wet, rocky/stony places with abundance of LHPs and robust, nectar-rich flowers such as knapweeds and thistles, upon which both sexes are very fond of feeding, resting, or roosting. LHPs: principally white stonecrop (*Sedum album*).

Conservation • Although protected by European legislation, it appears to be under no threat apart from potential for damage to, or destruction of, its unprotected habitats.

Parnassius phoebus Small Apollo

Distribution • C Alps of France, S Switzerland, N Italy, and S Austria. 1600–2800 m, generally 1800–2200 m. Sporadic, very local, sometimes common.

Description & variation • Resembles *P. apollo*. Male upf with red pd spot in s8. Antennal shaft whitish, narrowly ringed dark grey (cf. *P. apollo*). All markings regionally variable. § E Switzerland. Ups dark markings sometimes more extensive, with uph red ocelli joined by black bar.

Flight-period • Late June to late August in one brood.

Habitat • Damp, grassy places, often near streams. LHP: yellow mountain saxifrage (*Saxifraga aizoides*).

♂ E Switzerland

Parnassius mnemosyne **Clouded Apollo**

Distribution • Pyrenees through S France, C Alps, Italy, &
N Sicily to S Poland, S Fennoscandia (except Denmark),
Greece, & European Turkey. 75–2300 m, generally
1000–1700 m. Sporadic, often locally common.

♂ N Greece

Description & variation • § C & S
Greece. Upf with 4–6 white pd spots in
apical area; occurs as a variant in most
S European populations.

Flight-period • Mid-April to late August
in one brood. Emergence date depends on
altitude.

Habitat • Dry or damp, light woodland,
bushy places, or open, rocky, and grassy
slopes/gullies. LHPs: corydalis, including
solid-tubered corydalis (*Corydalis solida*),
bulbous corydalis (*C. bulbosa*), &
C. intermedia.

♀ N Greece

♀ N Greece

Pieridae

Well-represented in Europe, this family contains several of the more common and easily recognized butterflies, as well as some of the best known migrants, for example the Bath White, Small White, and Clouded Yellow. Mostly the differences between sexes are well-marked. Whilst many species are powerful fliers, especially the 'clouded yellows', others, for example the 'wood-whites', are strikingly weak. Without exception, the European examples of this family never sit with fully open wings: most species rest with partially open or half-opened wings, whilst the 'brimstones', 'clouded yellows', and 'wood-whites' only very rarely, if ever, open their wings at rest.

Aporia crataegi **Black-veined White**

Distribution • Most of Europe, including Sicily, Kefalonia, & several E Aegean Islands. Absent from Atlantic Islands, Balearic Islands, Corsica, Sardinia, Crete, & W Aegean Islands. 0–2000 m. Extinct in British Isles. Widespread, generally common.

Description • Ups & uns white, with distinctive black veins – unlikely to be confused with any other species.

♀ NW Greece

Flight-period • One brood. N Europe: late May to early July. S Europe: mid-April to July.

Habitat • Warm, sunny, bushy places; cultivated areas, especially orchards. LHPs: blackthorn (*Prunus spinosa*) & most cultivated members of plum family, including plum (*P. domestica*), peach (*P. persica*), apricot (*P. armeniaca*), & almond (*P. amygdalus*); also hawthorn (*Crataegus monogyna*), pear (*Pyrus communis*), apple (*Malus domestica*), & mountain ash (*Sorbus aucuparia*).

Behaviour • Strongly attracted to nectar of robust plants such as knapweeds & thistles. Shows marked dispersive tendency, following emergence, pairing, & feeding: specimens are often found on barren ground at high altitude, well removed from breeding grounds.

♀ NW Greece

Pieris brassicae Large White

Distribution • Throughout Europe, except Azores & Canary Islands. Possibly extinct in Madeira. 0–2600 m; records for highest altitudes probably relate to migrants. Widespread & common.

Description & variation • First brood: unh dusted with dark scales; male upf dark apical patch extending along outer margin to v3 or v2 (cf. *A rapae*). Second brood: unh dark colouring reduced or absent; male upf apical patch intensely black.

Flight-period • March to late October in several broods.

Habitat • Most habitat types containing LHPs, which include several members of cabbage family (Brassicaceae); also garden nasturtium ((*Tropaeolum majus*) (Tropaeolaceae)). Common garden pest on cabbages & nasturtium.

Behaviour • Powerful migrant. Vagrant specimens not uncommon on barren ground at high altitudes, well-removed from breeding ground. Common visitor to nectar-rich plants such as thistles, knapweeds, & 'butterfly-bush' (*Buddleia davidii*).

♀ S Greece (1st brood)

♀ S Greece (1st brood)

♂ SW France (2nd brood)

Pieris cheiranthi
Canary Islands' Large White

Distribution · Canary Islands (La Palma, widespread, locally common. N Tenerife). 200–1400 m.

Description · Resembles *P. brassicae* but ups & uns heavily marked, with larger upf pd spots often united. (*P. brassicae* not known to occur in Canary Islands.)

Flight-period · Throughout year in several broods.

Habitat · Usually wet, shaded places in laurel forests. LHPs: *Crambe strigosa* (Brassicaceae) & garden nasturtium ((*Tropaeolum majus*) (Tropaeolaceae)). (Unlike *P. brassicae*, does not frequent dry, cultivated areas under cultivation.)

Behaviour · Unlike *P. brassicae*, shows no migratory tendency.

Conservation · Destruction of natural habitat appears largely responsible for decline in many areas, & possibly extinction on some islands.

Artogeia rapae **Small White**

Distribution · Throughout Europe, including Azores, Madeira, Canary Islands, & Mediterranean islands. Generally very common, although less frequent in N Fennoscandia where occurrence may depend more on migration. 0–3000 m; records at highest altitudes more probably relate to migrants.

Description · First brood. Both sexes: upf apical patch extends along outer margin to v7 or v6 (cf. *A. mannii*). Male ups gc white; upf round, grey spot in s3 usually present; uph dark spot on costa variable. Female ups markings similar but heavier, with additional dark, elongate spot in s1b. Second brood. Markings bolder.

♀ NE Greece (2nd brood)

Flight-period · Generally early March to November in several broods. Recorded in all months in Canary Islands.

Habitat · Almost any ground containing LHPs, typically many members of cabbage family (Brassicaceae); also caper ((*Capparis spinosa*) (Capparidaceae)) & garden nasturtium ((*Tropaeolum majus*) (Tropaeolaceae)).

Behaviour · Probably most successful of European migrant butterflies. Common pest on cultivated cabbages.

Artogeia mannii **Southern Small White**

Distribution • Spain (very sporadic & local), through SW & S France, S Switzerland, & Italy (including Sicily), to Balkans & Greece, including Samos. 0–2000 m. Generally sporadic & local.

Description • Resembles *A. rapae* but markings generally heavier. Upf apex black, extending along outer margin to v4 or v3; distal margin of spot in s3 concave or linear (not round), often linked to outer margin by black scaling along v3 & v4; uph distal margin of costal mark concave.

Flight-period • March–September in 2+ broods.

Habitat • Dry, usually hot, rocky places, often in open scrubland or woodland. LHPs: evergreen candytuft (*Iberis sempervirens*) & rock candytuft (*I. saxatilis*).

Behaviour • Males often drink from damp soil.

♀ NW Greece (2nd brood)

♀ NW Greece (2nd brood)

Artogeia ergane **Mountain Small White**

Distribution • Spain through S France to Italy, Austria, Balkans, & Greece. 75–1850 m, generally above 500 m. Very sporadic & local in western range; widespread, locally common in SE Europe.

Description • Male upf without black spot in s3; apical patch somewhat square (cf. *P. rapae*). Female ups creamy, usually with extensive greyish suffusion; ups spots in s1b & s3 prominent. Second brood: ups markings heavier.

♂ S Greece (2nd brood)

Flight-period • Early April to late August in 2–3 broods.

Habitat • Hot, dry, bushy, rocky, or grassy places. LHP: principally burnt candytuft (*Aethionema saxatile*).

Behaviour • Males often assemble in large numbers on damp soil to drink.

Artogeia napi **Green-veined White**

Distribution • Most of Europe, including Mediterranean islands of Corsica, Elba, Sicily, Corfu, Thassos, & Lesbos. Generally widespread & common; very sporadic in S Spain & S Greece. Absent from Atlantic islands, Shetland Islands, & Sardinia. 0–2000 m.

Description & variation • Ups gc white. First brood: ups veins suffused light grey, unh yellow, veins suffused grey, giving an overall yellowish-green appearance. Male upf apical black scaling on veins variable; spot in s3 variable, sometimes absent. Female ups & uns black markings better developed, with additional spot in s1b & s5 upf. Summer broods in both sexes:

♀ NE Greece (2nd brood)

ups grey suffusion reduced, but usually with black scaling on veins near wing-edge; unh paler yellow, sometimes almost white (giving rise to possible confusion with *A. rapae*). Regionally extremely variable. Male ups greyish suffusion sometimes extensive. Female ups gc yellowish to greyish brown with variable greyish suffusion. Both sexes: unh deep yellow or yellowish green with heavy grey suffusion, or white with little or no suffusion. Darker forms generally occur in cooler climates, e.g., in Scotland, Scandinavia. Female upf with greyish line in s1b connecting outer margin to pd mark – so-called *bryoniae*-streak – occurs occasionally in England, commonly in Scotland, & is typical in N Scandinavia (cf. *A. bryoniae*).

♀ NE Greece (2nd brood)

♂ NE Greece (1st brood)

Flight-period • Number of broods & emergence date depends on region altitude & seasonal weather conditions. Lappland: June–July in one brood. England: April–May & mid-June to July in two broods. Further south: March–October in 2–4 partially overlapping broods.

Habitat • Damp, shady, grassy, & flowery places, often wooded river valleys in Mediterranean region. LHPs include watercress (*Nasturtium officinale*), lady's smock (*Cardamine pratensis*), jack-by-the-hedge (*Alliaria petiolata*), tower rock-cress (*Arabis turrita*), perennial honesty (*Lunaria rediviva*), & annual honesty (*L. annua*).

♂ N Norway

Artogeia balcana
Balkan Green-veined White

Distribution • S Balkans & Greece. 300–900 m. Distributional detail uncertain owing to probable confusion with *A. napi* unknown.

Description • Resembles *A. napi* very closely: unh veins more poorly defined in summer broods. Intermediate forms reported from Greece.

Flight-period • Early April to October in 2–3 broods.

Habitat • As for *A. napi*.

Artogeia bryoniae
Mountain Green-veined White

Distribution • From Jura Mts through C Alps & Tatra Mts to Carpathian Mts. 800–2700 m. Widespread, locally common.

Description • Male resembles first brood *A. napi*. Female ups variable; pale creamy yellow, suffused grey or greyish brown; upf with narrow, greyish streak in s1b connecting outer margin to pd mark – generally absent in *A. napi*.

Flight-period • Generally mid-June to early August in one brood; occasionally two broods (August–September) at lower altitudes in S Switzerland.

Habitat • Alpine/subalpine meadows. LHPs include alpine bittercress (*Cardamine bellidifolia*), mignonette-leaved bittercress (*C. resedifolia*), buckler mustard (*Biscutella laevigata*), alpine pennycress (*Thlaspi alpinum*), & mountain pennycress (*T. montanum*).

Artogeia krueperi Krueper's Small White

Distribution • S Balkans & Greece, including Corfu, Samos, Chios, & Kos. 75–1250 m, generally above 600 m. Sporadic & very local.

♀ (left) & ♂ S Greece (1st brood)

♀ S Greece (1st brood)

Description • Upf black apical patch extending along margin to v3 as small marginal triangles on each vein; black, wedge-shaped mark on costa near apex & large black pd spot in s3 conspicuous & distinctive; unh dark markings tending to greenish or yellowish colour, paler & less extensive in later broods.

Flight-period • Late March to late August in 3+ broods.

Habitat • Hot, dry, flowery, rocky, sometimes very steep slopes, usually on limestone. LHPs: golden alyssum (*Alyssum saxatile*) & mountain alyssum (*A. montanum*).

Behaviour • Both sexes appear to exploit air currents in patrolling rocky slopes of their habitat.

Pontia edusa **Eastern Bath White**

Distribution • NW France. E France, through S Switzerland, Italy (including Elba & Sicily), & E Germany (E of Rhine Valley), to S Fennoscandia, European Turkey, Greece, & most eastern Mediterranean islands. Recorded rarely as a migrant in S Ireland & S England. Occurrence in NE Europe largely depends on migration. Distribution in France & W Germany uncertain owing to possible confusion with *P. daplidice* & migratory behaviour of both species. 0–1500 m, with migrants to at least 2300 m. Widespread & generally common, especially in later broods.

Description & variation • Ups pattern of black markings very bold & distinctive, especially in female. Unh with complex pattern of largely confluent, dark green spots, tending to yellowish green in late broods; marginal spots somewhat rounded (cf. *P. chloridice* & *P. callidice*). Superficially indistinguishable from *P. daplidice*.

Flight-period • March to late October in 2+ broods. Emergence date depends on locality.

Habitat • Hot, dry, sometimes barren & stony places; common on disturbed ground, especially areas of cultivation. LHPs: principally white mignonette (*Reseda alba*), wild mignonette (*R. lutea*), & weld (*R. luteola*).

Behaviour • Well-known migrant.

♂ S Greece

Pontia daplidice **Bath White**

Distribution • Canary Islands. Madeira (very rare as migrant). Portugal & Spain (including Mallorca) to Corsica, Sardinia, E France, & W Germany (W of Rhine Valley). 0–2000 m; migrants recorded at 2900 m. Exact distribution & residential status uncertain in some regions owing to possible confusion with *P. edusa* & migratory behaviour of both species. Widespread & generally common, especially in later broods.

Description & variation • Superficially indistinguishable from *P. edusa*.

Flight-period • March–October in 2+ broods, but recorded in all months in Canary Islands.

Habitat & behaviour • As for *P. edusa*.

♂ S Spain

Pontia chloridice **Small Bath White**

Distribution · Republic of Macedonia. Bulgaria. NE Greece. European Turkey. 25–500 m. Extremely sporadic; very local, but usually abundant.

♂ NE Greece (2nd brood)

♀ NE Greece (2nd brood)

Description & variation · Male uph lacking black mark on costa; unh with marginal green to yellowish green stripes along veins (cf. *P. edusa* & *P. daplidice*). Second brood: uns green markings slightly paler, tending to yellow.

Flight-period · April–July in two prolonged, overlapping broods.

Habitat · Stony & gravelly places, including river-banks & dry river beds. LHP: *Cleome ornithopodioides* (Capparidaceae).

Behaviour · Roosts amongst stones or in rock crevices. Both sexes show some tendency to disperse in second brood, which usually is of much greater abundance.

Conservation · Grazing, gravel, & water extraction threaten most habitats associated with river systems.

♀ NE Greece (2nd brood)

Pontia callidice **Peak White**

Distribution • Pyrenees & C Alps. 1500–3400 m.

Description • Male ups white; upf with black bar at cell-end; subapex with black-lined veins in submargin near apex expanding into small black triangles at wing-edge; greyish pd markings variable, better developed near apex; unf colour & markings similar; unh largely yellow or yellowish green with heavy grey suffusion on veins uniting to give distinctive pattern of arrow-shaped or triangular white markings. Female ups pattern similar, with additional arrow-shaped marginal markings uph; all ups & uns markings heavier.

Flight-period • Early June to early August in one brood.

Habitat • Open, grassy, & rocky alpine slopes. LHPs: *Erysimum helveticum*, pyrenean mignonette (*Reseda glauca*), alpine bittercress (*Cardamine bellidifolia*), & chamois cress (*Hutchinsia alpina*).

Euchloe crameri **Western Dappled White**

Distribution • Portugal through S France to NW Italy. 0–2000 m. Widespread & common.

Description & variation • Resembles *E. ausonia* very closely. Ups gc white. First brood: ups discoidal spot usually not projecting along costal vein (cf. *E. simplonia*); hw apex (v8) conspicuously angled (cf. *E. tagis*). Second brood: larger; upf costa usually clear white – lacking dark striae; unf apex & unh markings yellowish green. No other member of group known to occur in same region.

♂ S Spain (1st brood)

Flight-period • March–July in two prolonged, overlapping broods.

Habitat • Open, hot, dry, flowery places; common on disturbed ground, especially in cultivated areas. LHPs: charlock (*Sinapis arvensis*), buckler mustard (*Biscutella laevigata*), wild radish (*Raphanus raphanistrum*), violet cabbage (*Moricandia arvensis*), evergreen candytuft (*Iberis sempervirens*), *I. pinnata*, & woad (*Isatis tinctoria*).

Euchloe simplonia
Mountain Dappled White

Distribution • N Spain (Cantabrian Mts & Pyrenees). Alps of SE France, SW Switzerland, & NW Italy. 400–2400 m, usually above 1500 m. Locally common.

Description • Ups gc white; upf black discoidal spot variable, usually narrow, curved, with short distal & proximal projections near costal vein (cf. *E. ausonia*; *E. crameri*). Hw apex (v8) conspicuously angled (cf. *E. tagis*).

Flight-period • April–August in one prolonged brood.

Habitat • Alpine/subalpine flowery meadows; grassy, rocky slopes. LHPs: buckler mustard (*Biscutella laevigata*), spoon-leaved candytuft (*Iberis spathulata*), & *Erucastrum nasturtiifolium*.

Euchloe ausonia Eastern Dappled White

Distribution • C & S Italy. Elba. Sicily. Balkans. Greece, including Crete, Rhodes, & several other Aegean islands. 0–1600 m. Widespread & common.

Description & variation • Resembles *E. crameri* & *E. simplonia* very closely, but geographical ranges not known to overlap. Variation between broods as for *E. crameri*.

Flight-period • Early March to early July in two prolonged, overlapping broods.

Habitat • Open, hot, dry, flowery places; common on disturbed ground, including cultivated areas. LHPs: charlock

(*Sinapis arvensis*), woad (*Isatis tinctoria*), *I. glauca*, burnt candytuft (*Aethionema saxatile*), evergreen candytuft (*Iberis sempervirens*), *Biscutella mollis*, buckler mustard (*B. laevigata*), crested bunias (*Bunias erucago*), & golden alyssum (*Alyssum saxatile*).

♀ NE Greece (1st brood)

♂ NE Greece (1st brood)　　　　　♂ NE Greece (1st brood)

Euchloe insularis **Corsican Dappled White**

Distribution • Corsica & Sardinia. 0–1300 m. Common & widespread.

Description • Hw apex (v8) conspicuously angled (cf. *E. tagis*); unh white spots small (cf. *E. ausonia* & *E. simplonia*). No other member of group occurs in same region.

Flight-period • Generally, mid-March to April & mid-May to late June in one brood. Partial second brood, generally of very low abundance, has been recorded in late summer to early autumn.

Habitat • Flowery scrub, rocky slopes, & gullies. LHPs include *Iberis pinnata* (Corsica), *Sinapis*, & hoary mustard (*Hirschfeldia incana*).

Euchloe tagis **Portuguese Dappled White**

Distribution • Portugal. Gibraltar. Spain. S France. NW Italy. 300–2400 m, generally below 1000 m. Extremely sporadic & local.

Description & variation • Ups markings resembles *E. ausonia*. Hw apex smoothly curved (cf. *E. ausonia*, *E. simplonia*, *E. insularis*, & *E. crameri*). Regionally variable, especially development of unh white markings.

Flight-period • Late March to May in one brood.

♀ S Spain

Habitat • Hot, dry, rocky, flowery slopes or gullies; sparse scrub, fallow fields, or margins of cultivation. LHPs: candytufts, including *Iberis pinnata*, burnt candytuft (*I. saxatilis*), evergreen candytuft (*I. sempervirens*), wild candytuft (*I. amara*), & *I. umbellata*.

Euchloe belemia **Green-striped White**

Distribution • Canary Islands (Fuerteventura, Tenerife, & Gran Canaria), 200–2300 m. S Portugal & Spain, 0–1350 m. Generally widespread & common.

Description & variation • Upf resembles *E. ausonia*; unh pattern of green repeated on uph as pale, grey marks; unf apical patch & unh green stripes dark, well-defined in first brood, tending towards a yellowish colour and somewhat diffuse in second brood. § Canary Islands. Smaller; upf black discal spot narrower.

♂ S Spain

Flight-period • Portugal & Spain: February to mid-April & late April to early June, in two overlapping broods. Canary Islands: late December to early June. Rarely, single specimens recorded well outside normal flight-times.

Habitat • Dry, flowery places amongst scrub or rocks, open woodland, or areas of cultivation. LHPs include buckler mustard (*Biscutella laevigata*), *Diplotaxis siifolia*, *Sisymbrium erysimoides*, *Descurainia bourgeana*, & *Carrichtera annua*.

Elphinstonia charlonia **Greenish Black-tip**

Distribution • Canary Islands (Fuerteventura, Lanzarote, & Graciosa, 0–400 m). Spain: Granada (Baza, 800 m) & Huesca (Fraga, 600–700 m). Extremely local, sometimes common.

Description • Resembles *E. penia* closely. Unf discal spot solid black; hair-collar between head & thorax rose pink.

Flight-period • Canary Islands: December–May in 2+ broods. Spain: late February to late May.

Habitat • Hot, arid, rocky places with very sparse vegetation. LHPs: Canary Islands, *Reseda lancerotae* & *Carrichtera annua*; Spain, *Eruca vesicaria* (plant of cultivated & disturbed ground).

Behaviour • Males show some tendency to wander, often appearing in areas far removed from breeding ground.

Elphinstonia penia
Eastern Greenish Black-tip

Distribution • S Balkans, European Turkey, & Greece. 700–1750 m. Very sporadic & extremely local.

Description • Resembles *E. charlonia* closely: upf discal spot faint, greyish; hair-collar between head & thorax yellow.

Flight-period • Early May to late July in two overlapping broods.

Habitat • Hot, dry, often precipitous limestone formations. LHP: tessellated stock (*Matthiola tessela*).

♂ N Greece

Behaviour • In hot, overcast conditions, both sexes often sit on warm rocks with half-open wings. Seeks shelter under rock-ledges in hottest part of day. Males show marked tendency to wander, often appearing at high altitude & in areas far removed from breeding ground.

Conservation • Many habitats vulnerable due to grazing of LHP; many colonies owe continued existence to precipitous nature of habitat.

♀ N Greece

Anthocharis cardamines **Orange Tip**

Distribution • Most of Europe, including Ireland, Britain, Baltic Islands, Corsica, Sardinia, Sicily, Corfu, Thassos, Lesbos, Chios, Samos, & Kastellorizo. Absent from Balearic Islands, Malta, Crete, & Rhodes. 0–2100 m. Widespread & common.

Description • Female upf apical patch dark grey. Both sexes: unf apex & unh mottled green to yellowish green.

Flight-period • Late March to June in one brood.

♂ N France

Habitat • Dry or damp meadows, marshes, lowland scrub, forest clearings, & alpine grassland. LHPs include lady's smock (*Cardamine pratensis*), hedge mustard (*Alliaria petiolata*), tower rock-cress (*Arabis turrita*), & honesty (*Lunaria annua*).

Anthocharis belia **Moroccan Orange Tip**

Distribution • Portugal & Spain, through S France to S Switzerland (very local), N & C Italy (very local). 0–1800 m. Widespread & common in Iberian Peninsula except NW Portugal.

Description & variation • Female upf orange apical patch with variable black suffusion; unf apex yellow. Both sexes: unh yellow with greyish-green basal & discal markings.

♂ S Spain

Flight-period • Generally April–June in one brood. Reported from Gibraltar & S Portugal (Algarve) in early March.

Habitat • Dry, often hot, flowery places, commonly margins of cultivated ground. LHPs: buckler mustard (*Biscutella laevigata*), *B. auriculata*, & *B. ambigua*.

Anthocharis damone Eastern Orange Tip

Distribution • S Italy, including Sicily (Mt Etna). Republic of Macedonia. C & S Greece, including Corfu (Mt Pantokrator). European Turkey. 350–1300 m. Very sporadic & local.

Description • Resembles *A. belia*; distributions do not overlap.

♂ S Greece

Flight-period • Early April to late May in one brood.

Habitat • Hot, rocky, often precipitous slopes on limestone. LHP: woad (*Isatis tinctoria*).

♂ S Greece

♀ S Greece

♀ S Greece

Anthocharis gruneri **Gruner's Orange Tip**

Distribution • Albania. Republic of Macedonia. SW
Bulgaria. Greece. 100–1800 m. Widespread but local.

♂ S Greece

♀ S Greece

Description • Male ups gc pale yellow.
Both sexes: unh white, mottled greyish
green.

Flight-period • Generally late March to
May in one brood. Emergence date depends
on locality: fresh males have been recorded
in early July.

Habitat • Hot, dry, rocky slopes, often
amongst scrub or small trees. LHPs: burnt
candytuft (*Aethionema saxatile*), less often
A. orbiculatum.

♀ S Greece

Zegris eupheme Sooty Orange Tip

Distribution • S & SW Spain to foothills of C Pyrenees. 500–1400 m. Generally very local.

Description • Upf black apex enclosing subapical, elongate, orange patch and preapical white mark on costa; orange patch variable, much reduced, and sometimes absent in female; black bar at cell-end prominent; upf gc whitish; uph gc creamy white.

Flight-period • March–June according to locality.

♂ C Spain

Habitat • Dry, flowery meadows; margins of cultivated land; neglected orchards, olive groves, etc. LHP: *Hirschfeldia incana*; woad (*Isatis tinctoria*).

Conservation • Threatened by herbicides, insecticides, & scrub clearance.

Colotis evagore Desert Orange Tip

Distribution • Coastal areas of S Spain. 0–400 m. Locally common.

♂ S Spain (2nd brood)

♂ S Spain (2nd brood)

Description • Ups black markings more extensive in female in first brood & in both sexes in later broods. Uns gc creamy white.

Flight-period • April–October in succession of broods.

Habitat • Hot, dry places, often near cultivated ground. LHP: caper (*Capparis spinosa*).

Behaviour • May disperse inland in late summer when butterflies often occur in great abundance.

Conservation • Habitat vulnerable due to close proximity to areas of intensive human activity. Rubbish dumping/burning & use of agricultural chemicals, or their improper disposal, is common threat to wildlife in coastal districts of S Spain.

♀ S Spain

Catopsilia florella **African Migrant**

Distribution • First reported from Canary Islands (Tenerife) in 1965; subsequently, from Gran Canaria (1966), Gomera, Fuerteventura, & Lanzarote (1976), La Palma (1986), & Hierro (1995). Recorded from Malta in 1963. Now resident in coastal districts; rarely recorded above 500 m. Local, sometimes abundant.

Description • Male ups very pale, greenish white; unh pale, dusky yellow. Female ups yellow or white; uns dusky yellow. Unlikely to be confused with any other species.

Flight-period • Throughout year in several broods.

Habitat • Flowery places, including gardens & parks. LHP: *Cassia didymobotrya* (an ornamental plant introduced from Africa).

Behaviour • Strongly migratory. LHPs are sometimes defoliated by an abundance of caterpillars.

Colias phicomone
Mountain Clouded Yellow

Distribution • Spain (Cantabrian Mts, above 1800 m). Pyrenees. C Alps. Romania (Carpathian Mts). 900–2500 m.

Description & variation • Resembles *C. nastes* but does not occur in same region. Male ups pale yellow, with heavy dark suffusion (variable); unh yellow, dusted with dark scales, submarginal band paler; white discal spot more prominent. Female similar, ups gc off-white.

Flight-period • Generally mid-June to mid-August in one brood, emerging late May at lowest altitudes.

Habitat • Grassy alpine meadows/slopes. LHPs: horseshoe vetch (*Hippocrepis comosa*), white clover (*Trifolium repens*), & bird's-foot trefoil (*Lotus corniculatus*).

Colias nastes **Pale Arctic Clouded Yellow**

Distribution • Lappland. 100–1100 m. Local, generally very common.

Description • Male ups gc variable, delicate shades of whitish yellow or green; marginal markings variable, light grey to black. Female similar, ups gc paler, markings heavier. Resembles *C. phicomone*, but geographically well-separated.

Flight-period • Mid-May to early July in one brood. Emergence date much dependent upon weather conditions.

♂ N Sweden

Habitat • Openheaths, grassy, or rocky places, often amongst light birch & willow scrub, usually near marshes. LHPs: alpine milk-vetch (*Astragalus alpinus*) & *Vaccinium*. NB: in most recent view, *C. nastes* is considered to be restricted to N America and Asia, European counterpart being regarded as *C. tyche*.

Colias palaeno **Moorland Clouded Yellow**

Distribution • France (Jura Mts & Vosges Mts), through C Alps & SE Germany to N Fennoscandia & N Balkans (Carpathian Mts). 100–2500 m. Extremely sporadic & local in southern & western range.

Description & variation • Male ups gc ranges from very pale yellow to pale sulphur yellow, with palest forms predominating in Lappland; unh yellow, heavily dusted with black scales giving an overall greenish appearance. Female ups normally white (sometimes pale lemon yellow in C Alps); unh slightly darker than male, somewhat yellowish.

Flight-period • June to late August in one brood. Emergence date depends on altitude, locality, &, in Arctic region, seasonal weather conditions.

♀ N Sweden

Habitat • Acidic marshes & peatbogs, typically with willow & birch scrub, & often near woodland. LHPs include northern bilberry (*Vaccinium uliginosum*) & bilberry (*V. myrtillus*).

Conservation • Habitat loss arising from land drainage responsible for widespread local extinction in many low-lying areas of C & NE Europe.

Colias chrysotheme **Lesser Clouded Yellow**

Distribution • E Austria. Slovakia. Hungary. Romania. 300–1000 m. Colonies very small, widely dispersed. Apparently extinct in Czech Republic.

Description • Male ups yellow veins in dark marginal borders distinctive (cf. *C. myrmidone*). Female fw pointed; upf costa dusky green; uph lemon yellow submarginal spots well-developed, extending to costa (cf. *C. crocea*).

Flight-period • Late April to late October in 2+ broods according to locality & weather conditions.

♂ W Hungary

Habitat • Grassy, flowery, bushy places. LHPs: hairy tare (*Vicia hirsuta*) & Austrian milk-vetch (*Astragalus austriacus*).

Conservation • Remaining colonies potentially very vulnerable due to small size & isolation. In general decline in western range.

Colias aurorina **Greek Clouded Yellow**

Distribution • Greece (Pindos Mts, Aroanian Mts, & Panahaikon Mts). 550–2000 m, generally below 1600 m.

Description • Male ups orange gc with very distinctive iridescent purple sheen, visible at oblique angles. Female ups

♂ S Greece

& uns gc white (commoner towards end of flight-period).

Flight-period • Mid-May to mid-July in one brood. Emergence depends on altitude & locality; rarely, worn examples may occur in late August.

Habitat • Hot, dry, rocky, & open grassy places with low scrub dominated by LHPs: *Astracantha rumelica* & *Astragalus parnassi* ssp. *cyllenus*.

♀ S Greece

Colias myrmidone Danube Clouded Yellow

Distribution • SE Germany to C Balkans. 100–500 m. Extremely sporadic & local.

Description • Resembles *C. caucasica* closely, but smaller. Male ups dark borders not crossed by yellow veins (cf. *C. chrysotheme*). Female gc rarely white, faintly tinged green. Does not occur in same region as *C. caucasica*.

Flight-period • Late May to mid-June, & mid-July to mid-September, in two broods.

Habitat • Open, bushy areas dominated by LHPs: *Cytisus ratisbonensis* & *C. capitatus*.

Conservation • In western range, all habitats extremely small & isolated. Declined in recent decades, possibly now extinct in Hungary.

Colias hecla Northern Clouded Yellow

Distribution • Lappland: Arctic Circle to Arctic Sea. 50–900 m. Very sporadic & local.

♂ N Norway

♀ N Norway

Description • Ups gc warm orange. Male ups borders black or dark brown. Female ups dark borders with yellowish spots. Female ups & uns gc sometimes very pale yellowish white.

Flight-period • Mid-June to early August in one brood. Emergence date much dependent on weather conditions.

Habitat • Open, grassy slopes with low-growing shrubs. LHP: alpine milk-vetch (*Astragalus alpinus*).

Colias hyale Pale Clouded Yellow

Distribution • Pyrenees to S Fennoscandia, through C Balkans & N Bulgaria. Migrants occur very rarely in Britain & S Norway, more often in N Germany & S Sweden. 0–1800 m. Generally sporadic, often locally common. Due to possible confusion with *C. alfacariensis*, records for Spain remain unconfirmed.

Description & variation • Resembles *C. alfacariensis* very closely: probably indistinguishable without close examination of both wing-surfaces. In general, ups & uns of *C. hyale* slightly paler, more lemon yellow. Female white gc sometimes replaced by yellow.

♀ N Spain

Flight-period • Early May to early October in 2–3 broods.

Habitat • Flowery, grassy places, usually associated with main LHP: lucerne (*Medicago sativa*) under cultivation.

Behaviour • Flight fast & powerful. Rests, roosts, & feeds with closed wings.

Colias crocea Clouded Yellow

Distribution • Most of Europe, including Canary Islands, Azores, Madeira, & Mediterranean islands, to S Scandinavia & Baltic countries. Appearance in Ireland, Britain, S Scandinavia, & Baltic countries depends exclusively on migration. 0–3200 m, generally below 1600 m. Very common & widespread in C & S Europe but probably not resident except in warmer regions of Mediterranean.

Description & variation • Male ups rich yellow gc sometimes replaced by pure yellow; dark marginal borders crossed by yellow veins. Female ups dark marginal borders enclosing yellow spots. § Female f. *helice*. Ups & uns gc sometimes replaced by pale, creamy white.

Flight-period • March–November in several broods in warmer Mediterranean region; throughout year in Canary Islands.

Habitat • Common in most habitat types, especially in warm–hot places with an abundance of flowers & LHPs, comprising many members of pea family (Fabaceae), most commonly lucerne (*Medicago sativa*), black medick (*M. lupulina*), red clover (*Trifolium pratense*), sainfoin (*Onobrychis viciifolia*), & white melilot (*Melilotus* alba).

Behaviour • Flight fast & powerful. Rests, roosts, & feeds with closed wings. One of commonest & best known European migrant butterflies.

♂ E Spain

Colias caucasica **Balkan Clouded Yellow**

Distribution • W & SW Balkans to NW Greece.
1200–2150 m. Very local & generally uncommon.

Description & variation • Resembles *C. myrmidone* but
larger. Male ups gc deeper orange. Female upf yellow spots in
black marginal border variable in size & number, sometimes
absent; gc rarely white. Does not occur in same region as
C. myrmidone.

Flight-period • Mid-June to mid-August in one brood.
Emergence date much dependent on altitude & weather
conditions.

Habitat • Alpine/subalpine grassland,
rocky slopes/gullies, or clearings in beech
woodland containing an abundance of
LHPs, two very distinctive species of broom:
Chamaecytisus hirsutus or *C. eriocarpus*.

Behaviour • Flight fast & powerful. Rests,
roosts, & feeds with closed wings.

♂ NW Greece

Colias alfacariensis
Berger's Clouded Yellow

Distribution • Most of C & S Europe, including Balearic
Islands, Corsica, & Sicily. Absent from Britain, N Holland, &
N Germany, except as rare migrant; also absent from NE &
S Greece, European Turkey, & Sardinia. 0–2100 m. Widespread
& very common in southern range; appearance in northern
range uncertain, depending on migration.

Description & variation • Resembles *C. hyale* very closely
– probably indistinguishable without close examination of
both wing-surfaces. In general, ups & uns of *C. alfacariensis*
slightly deeper ('less-pure') yellow. Female white ups & uns gc
sometimes replaced by yellow.

Flight-period • April–October in 2–3
broods.

Habitat • Dry grassy, often rocky places.
LHPs: horse-shoe vetch (*Hippocrepis comosa*)
& crown vetch (*Coronilla varia*).

Behaviour • Flight fast & powerful.
Rests, roosts, & feeds with closed wings.
Well-known migrant.

♀ S Spain

Colias erate **Eastern Pale Clouded Yellow**

Distribution • SE Europe, including N Greece & European Turkey. Distribution uncertain owing to migration & establishment of temporary colonies, especially in S Balkans. 0–1700 m. Very sporadic in western range.

Description & variation • Male ups gc pure lemon yellow; upf marginal borders usually solid black, but upf sometimes with yellow spots. Female ups, uns gc, & upf spots in marginal borders yellow or white. Sometimes hybridizes with *C. crocea*, giving rise to intermediate characters in gc & markings in both sexes.

Flight-period • Mid-March to October in 3–5 broods.

Habitat • Generally, hot, dry, flowery, grassy, & bushy places, mostly in areas containing an abundance of LHPs, principally lucerne (*Medicago sativa*) under cultivation, but also other species of *Medicago* (medick).

Behaviour • Flight fast & powerful. Rests, roosts, & feeds with closed wings. Well-known migrant.

Gonepteryx rhamni **Brimstone**

Distribution • Most of Europe, including Mediterranean islands of Corsica, Sardinia, Corfu, Kefalonia, & Zakynthos. Absent from Atlantic islands, Scotland, & N Fennoscandia. 0–2500 m. Widespread & common.

Description • Male ups uniform yellow (cf. *G. farinosa*). Female hw shape variable but dentation of inner margin better developed than that of *G. cleopatra*. Distinction from female *G. farinosa* often difficult: ups & uns white, faintly tinged green.

Flight-period • N & C Europe: June–July. S Europe: May–October. One brood, with hibernated specimens reappearing March to early May.

♂ S Greece

Habitat • Damp or dry bushy places or woodland margins. LHPs: buckthorns, including alder buckthorn (*Frangula alnus*), buckthorn (*Rhamnus catharticus*), Mediterranean buckthorn (*R. alaternus*), alpine buckthorn (*R. alpinus*), (*R. myrtifolia*), & olive buckthorn (*R. oleoides*).

Behaviour • On very warm days in early spring, males often break hibernation for brief periods. Both sexes often roost & hibernate amongst dense foliage of ivy, laurel, & other evergreen shrubs.

Gonepteryx cleopatra **Cleopatra**

Distribution • Iberia, including Balearic Islands. S & C
France: Pyrenees & Provençe to Vendée & Ain. Corsica.
Peninsular Italy except C Apennines; sporadic in northern
districts. Sardinia. Sicily. Dalmatian coast: sporadic. C & S
Greece, Corfu, Kefalonia, Zakynthos, Lesbos, Chios, Samos,
Ikaria, Rhodes, Karpathos, Crete, Skiathos, Skyros, Sifnos,
Paros, & Milos. European Turkey. Not reported from Albania
& Republic of Macedonia. Records for S Switzerland,
Bulgaria, & N Greece appear to relate to vagrant specimens
only. 0–1600 m, generally below 1200 m. Widespread, local.

Description & variation • Male ups gc yellow; upf with
bright orange discal patch; unf discal area yellow, unh & unf
costa pale yellowish green, whitish, or dull yellow, except for
slightly brighter upf discal area. Female ups gc white; unf with
delicate, pale orange streak above median vein; uph often with
faint orange flush; hw angular projection at v3 very shallow
(cf. female *G. rhamni* & *G. farinosa*). § Greece (including
Aegean Islands). Female ups gc sometimes
pale yellow or bright sulphur yellow
(resembling male *G. rhamni*). These forms,
including normal white form, may occur
together in widely varying ratios (usually
bright yellow on Rhodes).

Flight-period • Mid-May to August in
one brood, with hibernated specimens
reappearing late February to late April.

Habitat • Open, bushy, often rocky places,
sometimes in light woodland. LHPs:
buckthorns, including buckthorn (*Rhamnus
catharticus*) & Mediterranean buckthorn
(*R. alaternus*).

♂ E Spain

♀ E Spain

The following four species – the Atlantic island 'brimstones' – comprise a group closely related to *Gonepteryx cleopatra* (Cleopatra). Although easily recognizable as members of the very distinctive 'brimstone-group', their full taxonomic status has only recently been resolved by DNA analyses. Indeed, certainty regarding the character of *Gonepteryx maderensis* (Madeira Brimstone) remains unconfirmed, but due to its very distinctive character and the convenience of the present presentation, the butterfly is cited here as a distinct species. An interesting feature of these insects which will not be apparent, even to the most diligent butterfly-watcher, is their underside UV reflection patterns. It appears that members of this group which rest or roost amongst leaves of shrubs that are non-UV-reflective are themselves non-UV-reflective, in order to blend-in with the vegetation and so be less conspicuous to insectivorous birds which have very keen UV-eyesight. On islands where the practice of the butterfly is to rest or roost amongst leaves of shrubs which are UV-reflective, the underside of the butterfly is also UV-reflective, thereby effectively maintaining optimum camouflage.

Gonepteryx maderensis
Madeira Brimstone

Distribution • Known only from Madeira (Terreirro da Luta, Ribeiro Frio, & Encumeada Pass). 500–1500 m. Very local, not common.

Description • Male upf deep orange with very narrow, yellow marginal borders; fw & hw well-marked with a reddish brown marginal line, slightly expanded at veins. Female ups pale green with faint yellow flush; unh & unf costal areas dull green; unf disc whitish. Easily identified as no other brimstone butterflies occur on Madeira.

Flight-period • Recorded in all months but commoner April–September; number of broods uncertain, possibly only one, with adults hibernating erratically for short periods.

Habitat • Dense laurel (*Laurus laurocerasus*) forests containing other trees & shrubs including LHP: *Rhamnus glandulosa*.

Behaviour • Rests, roosts, & possibly hibernates amongst leaves of laurel.

Gonepteryx cleobule **Tenerife Brimstone**

Distribution • Known only from Tenerife (restricted to NE region – Anaga massif). 500–2000 m. Uncommon.

Description • Fw outer margin almost linear; hw dentation very shallow; fw & hw with conspicuous, reddish-brown marginal line, slightly expanded at veins. Male upf gc deep orange, extending almost to margins; uph yellow; unf greenish yellow; unh greenish.

Flight-period • Recorded in all months; number of broods uncertain, possibly only one, with adults hibernating erratically for short periods according to weather conditions.

Habitat & behaviour • Sunny clearings in laurel forests. Both sexes tend to fly near lower altitudinal range in winter & often take nectar from *Cedronella canariensis*. LHPs: *Rhamnus glandulosa* & *R. crenulata*.

Conservation • Becoming increasingly uncommon. Changes in restricted distribution & scarcity have been attributed to climatic change (increasing temperature, possibly due to continental drift towards Africa). Species disadvantaged by natural forces are all the more vulnerable to 'unnatural' forces – human interference with habitats.

♂ Tenerife, Canary Isles

Gonepteryx palmae **La Palma Brimstone**

Distribution • Known only from La Palma. 300–1600 m. Widespread, local, & uncommon.

Description • Fw outer margin almost linear; hw dentation very shallow. Male upf yellow with diffuse orange discal flush, variable; uph yellow; unh & unf costal area greenish yellow. Female ups very pale yellow – fw discal area almost white, hw & fw costal areas flushed pale yellowy-orange; uns similar. Easily identified as no other brimstone butterflies occur on La Palma.

Flight-period • Recorded March–April, June–September, & December. Number of broods uncertain, possibly only one, with adults hibernating erratically for short periods.

Habitat & behaviour • As for *G. cleobule*.

Gonepteryx eversi **Gomera Brimstone**

Distribution • Known only from Gomera. 500–1400 m. Local, uncommon.

Description & variation • Resembles *G. cleobule*. Upf orange discal flush slightly paler, variable. Female resembles *G. palmae*: ups lemon yellow; upf often with strong orange discal flush. Easily identified as no other brimstone butterflies occur on Gomera.

Flight-period • Recorded March–May, July–September, & December. Number of broods uncertain, possibly only one, with adults hibernating erratically for short periods.

Habitat & behaviour • As for *G. cleobule*.

Gonepteryx farinosa **Powdered Brimstone**

Distribution • S Balkans & Greece, including Levkas, Kefalonia, Rhodes, & Kastellorizo. 25–1450 m. Very sporadic & local, especially in N Greece.

Description • Male upf lemon yellow, with powdery-white appearance; uph distinctly paler – colour difference noticeable in flight. Female ups white, sometimes faintly tinged blue; unh dusky pale yellow (cf. *G. rhamni*).

♀ S Greece

Flight-period • Mid-June to July in one brood, with hibernated specimens reappearing March–April.

Habitat • Hot, dry, bushy, or rocky places. LHPs: buckthorns, including alpine buckthorn (*Rhamnus alpinus*), Sibthorp's buckthorn (*Rhamnus sibthorpianus*), & Christ's thorn (*Paliurus spina-christi*).

Behaviour • Rests & feeds with closed wings. Often roosts amongst leaves of Jerusalem sage (*Phlomis fruticosa*), LHPs, or other densely foliated bushes.

Leptidea sinapis Wood White

Distribution • Most of Europe, including Ireland, S Wales, S England, & most larger Mediterranean islands. 0–2300 m, generally below 1900 m. Widespread & common.

Description • Ups gc white; uns of black antennal club with small, white patch, extreme tip brown (cf. *L. duponcheli*). In later broods, dark markings reduced. Resembles *L. reali* very closely.

Flight-period • NE Europe: June to early August, generally in one brood. Most of C Europe: May–June & July–August, in two broods. S Europe: late March to September, often in three broods.

♂ N Greece (1st brood)

Habitat • Very wide-ranging in character. Woodland, scrubland, open flowery meadows; less often, sheltered grassy places above tree-line. LHPs include meadow vetchling (*Lathyrus pratensis*), bitter vetchling (*L. montanus*), bird's-foot trefoil (*Lotus corniculatus*), & greater bird's-foot trefoil (*L. uliginosus* – commonly used in Britain).

Behaviour • Flight weak & 'flappy'. Rests, roosts, & feeds with wings closed.

Leptidea reali Réal's Wood White

Distribution • Spain to SE Sweden & Balkans. 100–2000 m. Distribution and abundance uncertain due to probable confusion with *L. sinapis*.

Description • Resembles *L. sinapis* very closely. Black markings *averagely* denser, but differences insufficient for reliable identification. In both sexes, two species easily separated on basis of genitalia.

Flight-period & habitat • As for *L. sinapis*, with which it is usually found. LHP: meadow vetchling (*Lathyrus pratensis*).

Behaviour • Flight weak & 'flappy'. Rests, roosts, & feeds with wings closed.

Leptidea duponcheli **Eastern Wood White**

Distribution • SE France, NW Italy, S Balkans, N & C Greece, including Zakynthos. 50–1150 m. Sporadic, especially in Bulgaria, but locally common.

Description • Uns of antennal club brown (white in *L. sinapis*). Ups gc white. First brood: colour & pattern of dark grey uns markings distinctive, showing through to ups, giving an equally characteristic faint yellow or greenish appearance. Second brood: markings much less intense.

Flight-period • Mid-April to mid-May & late June to July, in two broods.

♂ (right) & ♀ NW Greece (1st brood)

Habitat • Hot, dry, often rocky places amongst scrub or in sparse woodland. Habitats generally smaller & averagely hotter/drier than those of *L. sinapis* with which it usually flies. LHPs: yellow vetchling (*Lathyrus aphaca*), meadow vetchling (*L. pratense*), & greater bird's-foot trefoil (*Lotus uliginosus*).

Behaviour • Flight weak & 'flappy'. Rests, roosts, & feeds with wings closed.

Leptidea morsei **Fenton's Wood White**

Distribution • S Poland. Slovakia. SE Austria. Hungary. N Croatia. Romania. Bulgaria. 250–1400 m. Very sporadic & local, especially in western range.

Description • Resembles *L. sinapis*. Fw apex distinctly 'hooked' – rounded in *L. sinapis* & *L. reali*. First brood: unh pattern & tone of greyish markings distinctive. Second brood: often much larger; markings greatly reduced. Antennal club resembles *L. sinapis*.

Flight-period • April to mid-May & mid-June to late July in two broods.

Habitat • Mature, usually damp, deciduous woodland margins or clearings. LHPs: spring vetch (*Lathyrus verna*) & black pea (*L. niger*). Habitat & LHPs often shared with *Neptis sappho*.

Behaviour • Flight weak & 'flappy'. Rests, roosts, & feeds with wings closed.

Lycaenidae

This family of generally small butterflies includes the 'Blues', 'Coppers', and 'Hairstreaks', and is represented by over one hundred species in Europe. For most species, sexual differences are well-marked, and the females of many 'blue' butterflies have brown uppersides, often with distinctive orange spots near wing-margins. On the other hand, close similarities between females of different but related species are sometimes less obvious, and closer inspection of diagnostic features is required for reliable identification. Some difficulties may also arise for the distinctive group known as the 'Anomalous Blues', so-called because the uppersides of both sexes are brown. Here, known distribution is often a useful aid in identification, and sexual distinction is made easy by the presence of a sex-brand on the upper fore-wing of the male.

The males of many species will often gather on damp ground to drink, sometimes in huge numbers and for prolonged periods, at which times identification may be attempted in a much more leisurely fashion; it may also provide a valuable opportunity to examine and compare several species at the same time. Flight is generally swift, and, for a few species, sometimes a little difficult to follow. The Long-tailed Blue and Short-tailed Blue are well-known migrants. For the dual purpose of locating habitats and identifying the butterflies themselves, a knowledge of larval host-plants is often invaluable. It is useful to note that host-plants of many Lycaenidae belong to the pea family.

Thecla betulae **Brown Hairstreak**

Distribution • Most of Europe, including W Ireland & S England. 50–1500 m. Very sporadic & generally extremely local.

Description • Male ups medium brown; upf usually with a pale yellow mark at cell-end; uph 'tails' orange. Female ups medium brown; upf with bright orange subapical band; uph orange 'tails' better developed.

Flight-period • Late July to early September in one brood.

♀ N Greece

Habitat • Deciduous woodland or mature scrub with open, sunny clearings containing an abundance of LHP: blackthorn (*Prunus spinosa*).

Behaviour • Adults are generally very secretive, often resting amongst leaves on higher bushes or trees.

Quercusia quercus **Purple Hairstreak**

Distribution • Most of Europe, including Britain. Reported from Sicily, Corsica, Sardinia, Crete, Lesbos, Samos, & Rhodes. 0–2100 m. Widespread, locally common.

Description & variation • Male ups with striking purple sheen visible from oblique angles; marginal borders dark grey. Female ups gc dark grey, with bright blue or bluish-purple patches in cell & inner margin. § Portugal, Spain, & S France. Uns paler, all markings less conspicuous.

♀ NE Greece

Flight-period • June–September in one brood.

Habitat • Hot, dry scrub, damp to dry woodland containing an abundance of LHPs: oaks (*Quercus*), including common oak (*Q. robur*); holly oak (*Q. coccifera*); holm oak (*Q. ilex*), & white oak (*Q. pubescens*).

Behaviour • Adults show little interest in nectar of plants, carrion, or excrement, & only rarely visit damp soil; nutrients seem to be obtained largely from aphid secretion ('honey-dew') on leaves of trees, especially LHPs, to which adult activity is largely confined.

♂ N Greece

Laeosopis roboris
Spanish Purple Hairstreak

Distribution • N Portugal, most of Spain through
E Pyrenees to SE France. 100–1600 m. Very sporadic, often
locally abundant.

Description • Ups gc greyish brown with extensive purple
flush on upf & uph, sometimes greatly reduced or absent in
female uph.

Flight-period • Late May to late July in one brood.
Emergence date depends on locality & local weather
conditions.

♂ S Spain

Habitat • Open, flowery, & bushy places
with LHP: ash trees (*Fraxinus excelsior*).

Behaviour • In early morning, both sexes
often assemble in large numbers to feed on
tall umbellifers (Apiaceae), including fennel
(*Foeniculum vulgare*). Flight is confined
mostly to higher branches of LHP in cooler
conditions, late afternoon to early evening.
Adults rest on leaves of trees & shrubs
during hottest periods.

Satyrium acaciae **Sloe Hairstreak**

Distribution • N & E Spain, to C E & SE Europe. 0–2000 m.
Generally widespread & common.

Description • Ups brown. Male ups without sex-brand.
Female abdomen with black anal hair-tuft.

Flight-period • June–July in one brood.

♂ NW Greece

Habitat • Dry or damp scrub, open
woodland, sheltered rocky gullies above tree-
line. LHP: blackthorn (*Prunus spinosa*).

Behaviour • Both sexes strongly attracted
to flowers of *Achillea* & thyme (*Thymus*).

Satyrium ilicis **Ilex Hairstreak**

Distribution • Portugal & Spain (including Balearic Islands) to S Fennoscandia, European Turkey, & Greece, including Corfu & many Aegean islands. 0–1600 m. Sporadic in western range, generally common.

Description & variation • Ups dark brown. Female upf often with orange suffusion in submarginal band. Male upf without sex-brand. § Portugal, Spain, & S France: upf with variable orange discal patch, usually better developed in female.

♂ S Greece

Flight-period • Late May to early August in one brood.

Habitat • Dry or damp scrub, light woodland, or heathland containing an abundance of LHPs: oaks (*Quercus*), including common oak (*Q. robur*), holm oak (*Q. ilex*), & holly oak (*Q. coccifera*).

Behaviour • Attracted to flowers of thyme (*Thymus*) & dwarf elder (*Sambucus ebulus*).

Satyrium esculi **False Ilex Hairstreak**

Distribution • Most of Portugal & Spain, including Balearic Islands (Ibiza & Mallorca). France (E Pyrenees to Maritime Alps). 500–1300 m. Sporadic & local.

Description & variation • Ups light greyish brown to dark brown; dull yellowy orange suffusion extremely variable, often absent. Male upf without sex-brand.

♀ S Spain

Flight-period • Late May to August in one brood.

Habitat • Hot, dry, flowery scrub or sparse woodland. LHPs: holm oak (*Quercus ilex*) & holly oak (*Q. coccifera*).

Satyrium spini **Blue-spot Hairstreak**

Distribution • From Portugal to C & SE Europe. Absent from Mediterranean islands, except Mallorca, Corfu, Zakynthos, Kithira, Limnos, & Lesbos. 0–2000 m. Generally widespread & common.

Description & variation • Ups brown; unh with distinctive blue spot in anal angle. Male upf with sex-brand. § Iberian peninsula. Female ups with variable orange suffusion, sometimes very extensive.

♀ NW Greece

Flight-period • Late May to late July in one brood.

Habitat • Hot, dry, grassy, & bushy places. LHPs include Mediterranean buckthorn (*Rhamnus alaternus*), alpine buckthorn (*R. alpinus*), & Christ's thorn (*Paliurus spina-christi*).

Satyrium w-album **White-letter Hairstreak**

Distribution • N Spain, Italy, & Greece (in Peloponnese, recorded only from Mt Chelmos), to S England & S Fennoscandia. Absent from European Turkey & Mediterranean islands except Sicily. 100–1300 m. Extremely sporadic & local.

Description • Ups brown. Male upf with sex-brand.

Flight-period • Generally mid-June to late July. Emergence may be delayed until August in S Scandinavia according to weather conditions.

♀ N Greece

Habitat • Mature woodland with open, sunny clearings. LHPs: elm, especially wych elm (*Ulmus glabra*). In last two decades, much habitat lost due to Dutch elm disease.

Behaviour • Both sexes strongly attracted to bramble blossom; rest for prolonged periods on leaves on higher branches of LHP.

Satyrium pruni **Black Hairstreak**

Distribution • Pyrenees, N Italy, & N Greece, to Denmark, S Sweden, & S Finland. C S England. 200–750 m. Extremely sporadic & local.

Description • Ups brown; ups submarginal orange markings variable; better developed in female. Male upf with sex-brand.

Flight-period • June to late July in one brood.

Habitat • Mature blackthorn (LHP: *Prunus spinosa*) thickets in sheltered, sunny clearings or at margins of mature, deciduous, predominantly oak woodland.

Behaviour • Adults take nectar from blossom of shrubs, especially privet (*Lingustrum vulgare*) & bramble (*Rubus fruticosus*) rather than low herbage. Often rest for prolonged periods in higher parts of LHP, so easily evading detection.

♀ N Greece

Conservation • Although damage to trees is often a matter of significant public concern, destruction of scrub, especially at woodland margins, receives relatively far less attention. Elimination of mature scrub, which may well contain plants older than many trees, may be of no less consequence than destruction of woodland itself. Many colonies of this butterfly, in company with many other species, have suffered from disregard or ignorance of this fact.

Satyrium ledereri
Orange-banded Hairstreak

Distribution • Greece: Samos (Mt Karvouni, Mt Kerketefs). 1000–1400 m.

Description • Ups brown.

Flight-period • June in one brood.

Habitat • Exposed, dry limestone slopes supporting sparse, low-growing vegetation. LHP: *Atraphaxis billardieri* (dock family (Polygonaceae)).

♀ Samos, Greece

Conservation • Population of Mt Karvouni seems potentially very vulnerable due to small size of habitat & easy accessibility.

Callophrys rubi **Green Hairstreak**

Distribution • Almost all of Europe, including Baltic Islands & most Mediterranean islands. Absent from Atlantic Islands, Orkney, Shetland, Outer Hebridian Islands, & Crete. 0–2300 m, generally below 2000 m. Widespread & common.

Description • Resembles *C. avis*. Frons green, eyes with very narrow white borders. Ups smoky brown, uns green. Unh with small white mark near costa, often with additional white dots on disc, sometimes developed into a dotted mediodiscal line on both wings.

Flight-period • March–June in one brood; fresh specimens occasionally recorded in July.

♀ NE Greece

Habitat • Well adapted to remarkable variety of habitat types, ranging from woodland clearings, bushy places, flowery meadows, heaths, & marshes, to sheltered rocky places or alpine grassland well above tree-line. Adaptation to wide range of LHPs no less remarkable: these include many varieties of broom (species of *Cytisus, Genista, Chamaespartium, & Chamaecytisus*), gorse (*Ulex europaeus*), kidney vetch (*Anthyllis vulneraria*), rockrose (*Helianthemum nummularium*), & flowers or fruits of several shrubs, including buckthorns (*Rhamnus*), bilberries (*Vaccinium*), dogwoods (*Cornus*), & bramble (*Rubus fruticosus*).

Behaviour • Rests, feeds, & roosts with closed wings.

Callophrys avis
Chapman's Green Hairstreak

Distribution • S & NW Portugal. S Spain (provinces of Cadiz, Malaga, Teruel, Gerona, & Barcelona). S France (SE Pyrenees to Maritime Alps). 100–1000 m.

Description • Resembles *C. rubi*. Frons & eye borders rusty red. Ups reddish brown, uns green. Unf & unh with thin white mediodiscal line.

Flight-period • Late March to mid-June in one brood.

Habitat • Dry scrub containing or dominated by main LHP: strawberry tree (*Arbutus unedo*).

Behaviour • Rests, feeds, & roosts with closed wings.

Conservation • Proximity to mostly highly populated coastal regions makes it potentially very vulnerable. Frequency of bush fires in S France is an additional threat.

Tomares ballus Provençal Hairstreak

Distribution • Iberia. France (SE Pyrenees to Maritime Alps). 300–1300 m. Sporadic, generally very local but sometimes common.

♂ S Spain

♀ S Spain

Description • Male ups medium brown; uns green or bluish green areas sometimes greatly reduced. Female ups gc light brown, with extensive orange discal patch on fw & in pd, posterior half of hw.

Flight-period • January to mid-May in one brood. Emergence date much dependent on locality & altitude (earliest in coastal districts of Algarve).

Habitat • Dry, flowery places, often with short grass/turf & sparse scrub. LHPs: principally medick, including bur medick (*Medicago minima*) & black medick (*M. lupulina*).

Behaviour • At high altitude, especially in early spring, adults often sit on or near rocks heated by early morning sun to hasten their recovery from overnight cold.

Tomares nogelii Nogel's Hairstreak

Distribution • SE Romania: known only from district of Dobrogea (information very limited).

Description • Ups resembles *T. ballus*; uns gc grey; upf & unh, with very distinctive submarginal, pd & discal orange spots edged with black marks.

Flight-period • May to early July in one brood.

Habitat • Grassy scrub. LHP: Pontic milk-vetch (*Astragalus ponticus*).

Lycaena helle **Violet Copper**

Distribution • S France (E Pyrenees) through Jura Mts, NW Switzerland, S Germany, S Belgium, & Poland, to Slovakia & Fennoscandia. 100–1800 m. Locally often very common in very small, widely dispersed colonies.

♂ S Belgium

♂ S Belgium

♀ NW Switzerland

prominent.

Flight-period • May–July in one brood.

Habitat • Flowery, marshy meadows & bogs, often by rivers or lakes, & commonly associated with woodland or scrub. LHPs: C Europe, common bistort (*Polygonum bistorta*); Fennoscandia, alpine bistort (*P. viviparum*).

Behaviour • Adults often bask with open wings on leaves of LHP, where this is common bistort, & frequently roost on tall flower stems of this same plant.

Conservation • Drainage & afforestation of habitats poses serious threat in most regions, except Fennoscandia.

Description • Male ups orange gc largely obscured by strong violet reflections; uph & unh orange submarginal band

Lycaena phlaeas Small Copper

Distribution • Canary Islands (Lanzarote). Madeira. Almost all of Europe including all larger & many smaller Mediterranean islands. Absent from Azores, Shetland, Orkney, & Outer Hebridian Islands. 0–2400 m. Widespread & common.

Description & variation • First brood: male ups brightly marked. Summer broods: ups often suffused smoky greyish brown, sometimes almost obscuring orange gc; hw 'tail' at v2 usually prominent. § North of Arctic Circle (0–400 m),

♂ (1st brood) NE Greece

♂ (2nd brood) NE Greece

L. p. polaris. Unh gc dove grey, with prominent black pd & discal spots, whitish pd striae, & conspicuous submarginal red spots in striking contrast. § Madeira (50–1800 m), *L. p. phlaeoides.* Ups resembles nominate form; unh gc greyish brown, irregular greyish pd band distinctive.

Flight-period • N Europe: May to early October in one or two broods. S Europe: February to late October in 3+ broods. Lanzarote: all year in succession of overlapping broods. Madeira: March–October in at least three broods.

Habitat • Adapted to most habitat types, ranging from very hot, arid, low-lying Mediterranean coastal regions to cool, damp, grassy, & flowery places of high mountains & Lappland. LHPs: dock family, especially sheep's sorrel (*Rumex acetosella*); in very hot, dry places, commonly knotweed (*Polygonum aviculare*).

♂ L.p. polaris (N Norway)

♂ (1st brood) NE Greece

Lycaena dispar Large Copper

Distribution · France. S Holland. Italy. Germany. Latvia. S Finland. Poland, through Balkans to N & C Greece & European Turkey. 0–1000 m. Colonies very few, very small, & widely dispersed in most of range; generally uncommon. Nominate form (*L. dispar dispar*) became extinct in about 1848 in fens of E England. Since 1927, a colony originating from Dutch race (*L. d. batava*) has been maintained at Woodwalton Fen, Huntingdonshire, England.

♂ N Greece

♂ N Greece

Description & variation · § Holland (Friesland), *L. d. batava*. Resembles nominate form very closely. § Rest of Europe, *L. d. rutila*. Resembles nominate form; generally smaller; unh variable, but usually duller, tending to yellowish grey; orange submarginal band paler. In SE Europe, individuals of second brood sometimes approach or exceed size of nominate form.

Flight-period · Holland & other northern regions: June–July in one brood. Elsewhere: late May to June & August, generally in two broods; occasionally, third brood has been reported from some localities in S Europe.

Habitat · Boggy margins of lakes, rivers, ditches, & canals. LHPs: great water-dock (*Rumex hydrolapathum*), curled dock (*R. crispus*), & water-dock (*R. aquaticus*).

Conservation · Changes in land use, particularly drainage of wetlands, pose serious threat; in Greece, particularly, all known colonies appear to be at imminent risk on this account.

♀ N Greece

Lycaena virgaureae **Scarce Copper**

Distribution • From C Spain through S France to Arctic Circle & C Greece. 1000–2000 m. Very sporadic & local.

Description & variation • § Lappland, f. *oranulus*. Male ups more yellow, black marginal borders sometimes slightly wider. Female ups suffused grey. § C Alps (1700–2000 m), *L. v. montanus*. Male ups black marginal borders wider, sometimes with very small black discal spot. Female ups dull golden yellow with variable greyish suffusion. § N & C Spain (600–1600 m), *L. v. miegii*. Male upf & uph with black cell-bar & pd spots, variable in number & size.

♀ N Greece

Flight-period • Late June to September in one brood. Emergence date depends on locality.

Habitat • Sheltered flowery places, often dampish clearings, or hill-side bogs in woodland. LHPs: docks, especially common sorrel (*Rumex acetosa*).

♂ N Greece   ♂ N Greece

Lycaena ottomana **Grecian Copper**

Distribution • S Balkans & Greece, including Corfu & Evia. 50–1500 m, generally below 1000 m. Widespread but often very local.

Description & variation • Second brood: hw 'tail' at v2 generally better developed but variable.

Flight-period • Mid-April to late May & July to early August in two broods.

Habitat • Dry, generally hot, grassy, flowery places, often amongst bushes or in light woodland. LHP: sheep's sorrel (*Rumex acetosella*).

Behaviour • Males of summer brood are especially attracted to flowers of dwarf elder (*Sambucus ebulus*). Females are secretive, giving impression of rarity even in colonies where males are common.

♀ NE Greece

♂ NE Greece

♂ NE Greece

Lycaena tityrus Sooty Copper

Distribution • From W & N Spain to C E & SE Europe to about 58°N in Lithuania. Absent from Mediterranean islands except Sicily & Samos. 50–2500 m. Widespread & common.

Description & variation • First brood: male ups dark greyish brown, contrasting strongly with brilliant white fringes; uph with small, orange submarginal spots. Female ups orange submarginal bands & orange basal suffusion extensive. Later broods: male uph orange submarginal spots better developed, sometimes extending to upf. Female ups brown, sometimes with orange discal suffusion; orange submarginal spots reduced. § Pyrenees (2100 m) & C Alps (1200–2500 m), *L. t. subalpinus*. Ups dark brown, submarginal orange spots much reduced or absent; uns gc yellowish grey. § C W Spain (900–1100 m), *L. t. bleusei*. Ups orange markings extensive, especially in female.

Flight-period • Hot, low-lying localities: mid-April to October in 3+ broods. Cooler localities: April–June & July–September in two broods. At highest altitudes: late June to September in one brood.

Habitat • Dry or damp, flowery & grassy places, often amongst bushes or at woodland margins; sheltered gullies at alpine levels. LHPs: docks, especially common sorrel (*Rumex acetosa*) & French sorrel (*R. scutatus*).

♂ NW Greece (2nd brood)

♂ N Greece (1st brood)

♂ NE Greece (1st brood)

♀ N Greece (2nd brood)

♀ NE Greece (2nd brood)

Lycaena alciphron **Purple-shot Copper**

Distribution • From Portugal to Estonia & European Turkey. 50–2500 m. Absent from Mediterranean islands except Sicily & Lesbos. Very sporadic & local.

♀ *L. a. melibaeus*, NW Greece

♀ *L. a. melibaeus*, NW Greece

Description & variation • Male ups gc orange, suffused violet grey. Female ups brown with variable orange markings. § N Portugal to S Switzerland & Italy, *L. a. gordius*. Male ups brighter. Female ups gc clear, bright orange, black markings heavy. § Balkans & Greece, *L. a. melibaeus*. Male ups smoky suffusion generally darker. Female ups brown, orange markings variable.

Flight-period • Generally June–August in one brood. Recorded from coastal districts of S France in late April.

Habitat • Grassy or rocky flowery places. LHPs: docks (*Rumex*).

Behaviour • Both sexes attracted to flowers of thyme (*Thymus*). Often basks with half-open wings.

♀ *L. a. gordius*, S Spain

♂ *L. a. melibaeus*, NW Greece

Lycaena thersamon **Lesser Fiery Copper**

Distribution • Widespread, generally very local.
C peninsular Italy, C E & SE Europe to S Peloponnese &
E Aegean islands of Thassos, Kos, & Rhodes. 0–1600 m.

Description & variation • Male uph posterior discal & pd
areas with slight greyish suffusion, sometimes tinged violet.

♂ NE Greece

Flight-period • April–October in 2+
broods.

Habitat • Dry, often very hot, flowery,
grassy, or rocky places. LHP: knotweed
(*Polygonum aviculare*).

Behaviour • Both sexes greatly attracted
to blossom of thyme (*Thymus*) & dwarf
elder (*Sambucus ebulus*).

♀ NE Greece

♀ NE Greece

Lycaena thetis **Fiery Copper**

Distribution • C & S Greece. 1500–2300 m. Very sporadic & local.

♂ S Greece

♀ S Greece

Description • Male unh colour slightly variable, almost pure powdery white to creamy off-white with very faint orange flush; dark discal & pd spots inconspicuous – somewhat 'ghosted'.

Flight-period • Mid-July to August in one brood.

Habitat • Generally exposed, dry, rocky sites on limestone supporting low-growing shrubs, including spiny, cushion-forming LHP: *Acantholimon androsaceum*. All known habitats in S Greece are shared with *Turanana endymion* (Odd-spot Blue).

Behaviour • Both sexes strongly attracted to nectar of thyme (*Thymus*).

Conservation • Exploitation of many higher mountains of Greece for recreational & other purposes poses direct threat to habitats.

♀ S Greece

Lycaena hippothoe **Purple-edged Copper**

Distribution • N Spain, through C France to N Fennoscandia & C E Balkans. 0–2500 m. Sporadic, often locally common.

Description & variation • Male ups with striking purple flush. Female upf orange discal area variable; uph orange submarginal markings prominent; unf orange discal flush variable. § C Alps, Apuane Alps, & C Apennines (1500–2500 m), *L. h. eurydame*. Male ups without purple flush. Female ups uniform brown, usually without submarginal orange markings, unf without orange flush. Altitudinal range overlaps that of nominate form but not on same mountains. § Fennoscandia from about 62°N to North Cape (0–400 m), *L. h. stiberi*. Male ups gc lighter golden red, unf orange discoidal flush & unh orange band distinct. Female upf discoidal area orange extending towards outer margin. Both

sexes: unh gc colour light greyish-buff; ups & unh submarginal bands well-developed.

Flight-period • W & N Europe generally: June to late July in one brood. High altitudes in C Alps: July to mid-September in one brood. C E Europe, including S & E Hungary: May & late July to August in two broods.

Habitat • Damp or wet alpine to subalpine meadows & hillside bogs; grassy places, often near shore-line on Norwegian coast. LHPs: bistort (*Polygonum bistorta*) & common sorrel (*Rumex acetosa*).

♂ S Belgium

Lycaena candens **Balkan Copper**

Distribution • S Balkans to N & S C Greece. 900–2000 m. Sporadic, locally common.

Description • Resembles *L. h. hippothoe* very closely; probably indistinguishable without reference to male genitalia.

Flight-period • Mid-June to late July in one brood.

♀ NW Greece

Habitat • Flowery hill-side bogs, often in woodland clearings; less often, exposed, dry, grassy places above tree-line. LHP: common sorrel (*Rumex acetosa*).

♀ NW Greece   ♂ NW Greece

Lampides boeticus **Long-tailed Blue**

Distribution • Canary Islands, Mediterranean Islands, &
Europe, to N Germany. Generally widespread & common in
southern regions. Occasionally common in Channel Islands.
Very rare in S England & generally very uncommon in
northern range, where occurrence probably depends solely on
migration. Residency uncertain in most regions, including
cooler parts of S Europe, where early spring broods probably
arise from N African migrants. (Residence possible only in
warmer parts of Mediterranean where continuity of larval
food source is maintained throughout winter). 0–2700 m,
generally below 1500 m.

Description • Male ups violet blue gc variable, reduced
scaling giving way to light brownish appearance especially near
wing-margins. Both sexes: uph dark anal spot in s2 prominent;
unh creamy pd band conspicuous (cf. *L. pirithous.*)

Flight-period • S Europe: February–November in several
broods. Northern range: date of appearance very uncertain &
depends solely on migration, itself greatly dependent upon
weather conditions. Canary Islands: all year, in overlapping
broods.

Habitat • Hot, dry, bushy, flowery places,
including margins of cultivated ground.
LHPs: many species of pea family
(Fabaceae); in S Europe, commonly bladder
senna (*Colutea arborescens*), broad-leaved
everlasting pea (*Lathyrus latifolius*), &
lucerne (*Medicago sativa*). Often a pest on
cultivated peas (*Psium sativum*) & broad
beans (*Phaseolus vulgaris*).

♀ S Greece

♂ S Spain

Cacyreus marshalli **Geranium Bronze**

Distribution • Early stages of this South African species believed to have been introduced accidentally to Mallorca in 1989 through importation of *Pelargonium* cultivars (geraniums), the LHP. Quickly became pest & has since spread to other Balearic Islands, Spanish mainland, & S France. Several colonies noted near Rome & male specimen captured in Brussels garden in 1991. So far, all records relate to villages & towns &, invariably, have been directly associated with occurrence of its very widespread & popular ornamental host plant.

Description • Ups dull, greyish brown with 2–3 small, dark spots in anal angle; hw with distinct tail at v8; fringes chequered black & white.

♀ Menorca, Spain

Leptotes pirithous **Lang's Short-tailed Blue**

Distribution • Mediterranean region, including islands. Less common in northern range where occurrence probably depends solely on migration. Residency uncertain in most regions, including S Europe, except in warmer parts where continuity of larval food source is maintained throughout winter. In relatively cooler, eastern Mediterranean region, early spring broods probably arise from N African migrants. 0–1200 m. Not recorded from Atlantic islands. Widespread & common.

Description • Male ups gc violet; dark marginal borders narrow. Both sexes: uph dark anal spots small, often obscure; uns marbled white & grey to greyish brown; unh lacking obvious pale pd band (cf. *L. boeticus*).

♀ Spain

Flight-period • Areas of residence:
February–October in several broods.
Northern range: date of appearance very
uncertain, depending solely on migration.

Habitat • Hot, dry, flowery places,
including scrubland & cultivated areas,
especially fields of lucerne (*Medicago sativa*)
which appears to be main LHP. Wide range
of other plant families have been recorded.

♀ Spain

Cyclyrius webbianus **Canary Islands' Blue**

Distribution • Known only from Canary Islands (Gomera,
La Palma, Tenerife, & Gran Canaria). Generally 200–2500 m,
but recorded to 3500 m on Mt Teide (Tenerife).

Description • Male ups dull blue with purple sheen visible
from oblique angles. Female ups brown, sometimes with
purple flush at base of fw.

Flight-period • Below 2000 m: throughout year in several
overlapping broods. Above 2000 m: not recorded between
early October & early May.

♀ Gomera, Canary Isles

Habitat • Below 2000 m: flowery, rocky
scrub. Above 2000 m: sheltered places with
sparse vegetation. LHPs: principally brooms
& trefoils, including *Cytisus canariensis* &
Lotus sessilifolius.

Behaviour • Adults strongly attracted to
nectar of *Micromeria* species.

Tarucus theophrastus **Common Tiger Blue**

Distribution • S Spain. Coastal districts from Cadiz to Murcia. 25–250 m. Extremely sporadic & very local.

Description • Resembles *T. balkanicus*. Uns black pd line disrupted by veins on both wings. The two species not known to occur together in Europe.

Flight-period • Spain: April–September in several broods. Often very scarce in first brood, but abundant in late summer.

Habitat • Hot, dry, open, quite often flat scrubland usually dominated by large bushes of LHP: common jujube (*Ziziphus lotus*).

♀ (right) & ♂ S Spain

Behaviour • During most of day, adults fly rapidly amongst branches or rest on leaves of LHP. In early evening, large numbers sometimes gather at tops of grass stems near LHP.

♀ S Spain

♂ S Spain

Tarucus balkanicus **Little Tiger Blue**

Distribution • Coastal district of eastern Adriatic Sea to European Turkey & N & C Greece, including Corfu. 50–850 m. Extremely sporadic & very local.

Description • Resembles *T. theophrastus*. Uns dark pd line usually unbroken by veins. The two species not known to occur together in Europe.

Flight-period • Mid-April to October in several broods. Very scarce in first brood, but population density increases rapidly during summer.

Habitat • Hot, dry, open, rocky scrub, usually dominated by very distinctive bushes of LHP: Christ's thorn (*Paliurus spina-christi*).

Behaviour • Both sexes warm themselves in early morning by sitting on rocks, usually with wings closed. Strongly attracted to nectar of small, purple-flowered *Micromeria*.

♀ N Greece   ♂ N Greece

Azanus ubaldus **Desert Babul Blue**

Distribution • Canary Islands (Gran Canaria: records limited to Playa del Inglés (1982) & Maspalomas (1992)).

Description • Male ups pale violet blue, with well-defined, slightly darker, blue androconial patch on upf; uns gc pale greyish brown, with transverse pale lines; unh with two conspicuous round spots on costa & two spots in anal angle. Female ups light brown, often with blue basal suffusion, otherwise similar to male.

Flight-period • Records relate to late January & late April, but known to be continuously brooded in N Africa.

Habitat & behaviour • Restricted to immediate vicinity of LHP: *Acacia* trees growing in very arid places. Details for Gran Canaria not known, but elsewhere in range, adults spend much time flying rapidly amongst branches of LHP, often taking nectar from flowers. Curiously, an entire colony may restrict itself to a single *Acacia* tree, despite presence of others nearby. Both sexes visit damp ground to drink.

Zizeeria knysna **African Grass Blue**

Distribution • Canary Islands. 0–100 m. Very local in coastal districts & apparently very rare on Gomera, Hierro, & Lanzarote. Spain: mainly coastal valleys in provinces of Cadiz, Malaga, Granada, & Almeria; apparently, no recent records from inland sites. Malta & Sicily. Recently confirmed for Crete, 25–800 m. Extremely sporadic & very local.

Description • Male ups dark, violet blue with variable black marginal borders. Female ups gc dark brown, sometimes with purplish-blue basal flush upf, less often extending to discal area of both wings. Both sexes: uns gc light greyish brown with marginal, pd & discal spots darker brown.

Flight-period • Canary Islands: all year in several overlapping broods. Spain: February–October in at least 3–4 broods.

♀ S Spain

Habitat • Generally very hot places, including coastal gullies, but usually near streams or springs. Inland sites comprise damp, sunny clearings in wooded river valleys. LHPs include amaranth (*Amaranthus*), bur medick (*Medicago minima*), & black medick (*M. lupulina*).

Conservation • Particularly threatened in Spanish coastal sites which invariably are close to centres of intense human activity.

Everes argiades **Short-tailed Blue**

Distribution • N Spain, through France, Switzerland, Italy, & Sicily to S Sweden (Gotland), S Finland to Balkans, & N Greece. In most of northern range, occurrence largely or entirely depends on migration. Very rarely recorded in S England. 0–1000 m. Generally very sporadic & local.

Description • Male ups violet blue; upf without discal spot; unh gc light grey with characteristic pd & discal white-ringed,

♀ NE Spain

black spots & orange spots in s1c & s2; hw 'tail' at v2 variable. Female ups dark, greyish brown, often with blue basal suffusion & small orange spot in anal angle; uns as for male.

Flight-period • April to mid-June & July–August in two broods.

Habitat • Flowery, grassy, & bushy places, or woodland clearings. LHPs include bird's-foot trefoil (*Lotus corniculatus*), crown vetch (*Coronilla varia*), black medick (*Medicago lupulina*), & wild liquorice (*Astragalus glycyphyllos*).

♂ N Greece

♂ N Greece

Everes decoloratus
Eastern Short-tailed Blue

Distribution • E Austria, through C Balkans to C N Greece. 250–1000m. Very sporadic & local, but sometimes common.

Description • Resembles *E. alcetas*. Male ups darker, dusky blue (density of blue scales extremely variable), black outer marginal borders narrow, indenting along veins; upf with small black discal spot. Female ups dark chocolate brown. Both sexes: unh lacking orange spots (cf. *E. argiades*).

♂ N Greece

Flight-period • May–June, July–August, & September in three broods.

Habitat • Flowery, bushy places; sunny clearings in light deciduous woodland. LHPs: black medick (*Medicago lupulina*) & lucerne (*M. sativa*).

Everes alcetas **Provençal Short-tailed Blue**

Distribution • N Spain & SW France, through Corsica, N Italy to C Balkans, & N & C Greece. 50–1200 m. Very sporadic, local, & generally uncommon.

Description • Resembles *E. decoloratus*. Male ups more violet blue, with very narrow black borders; upf without black discal spot. Female ups dark greyish brown. Both sexes: unh lacking orange spots (cf. *E. argiades*).

♂ NW Greece

Flight-period • Late May to June, July–August, & late September in three broods.

Habitat • Open, grassy, flowery places in light deciduous woodland. LHPs: crown vetch (*Coronilla varia*) & goat's rue (*Galega officinalis*).

♂ N Greece

♀ SW France

Cupido minimus **Little Blue**

Distribution • From N & E Spain through Italy, Ireland, Britain, & Lappland, to Balkans & Greece, including Corfu & Kos. 50–2800 m. Widespread, locally common.

Description • Male ups gc greyish to almost black, dusting of pale bluish scales variable, but always present at wing-base. Female ups dark greyish brown.

♂ NW Greece

♂ N Greece

Flight-period • April–July in one brood, or April–June & July–September in two broods, according to altitude & locality.

Habitat • Fairly dry grassy & flowery places, sometimes amongst rock or in woodland clearings. LHP: kidney vetch (*Anthyllis vulneraria*).

♀ SW France

Cupido carswelli **Carswell's Little Blue**

Distribution • Known only from mountains of S & SE Spain. 1000–1800 m.

Description • Resembles *C. minimus*, except male ups have a small patch of purple scales at wing-base, sometimes extending along some veins of hw.

♂ S Spain

Flight-period • Late April to early June in one brood.

Habitat • Hot, dry, rocky grassland, often amongst open scrub. LHP: dark red variety of kidney vetch (*Anthyllis vulneraria*).

♀ S Spain

Cupido osiris **Osiris Blue**

Distribution • From Spain to Balkans, European Turkey, & Greece, including Aegean island of Limnos. 500–1800 m. Sporadic in most of range, generally very local.

Description • Male ups violet blue with sharply defined, very narrow, black marginal borders. Female ups dark brown; upf sometimes with blue basal flush.

Flight-period • May–July in one brood, or April to late June & late July to early September in two broods, according to locality & altitude. Second brood may fail to appear in exceptionally dry conditions due to desiccation of LHP.

Habitat • Grassy, flowery, sometimes rocky places. LHPs: sainfoins, including sainfoin (*Onobrychis viciifolia*) & mountain sainfoin (*O. montana*).

♂ N Greece

♂ N Greece

Cupido lorquinii **Lorquin's Blue**

Distribution • S Portugal (Algarve). S Spain (provinces of Cadiz, Malaga, Granada, & Jaen). 100–2000 m. Very sporadic & local.

Description • Male ups violet blue with broad black marginal borders. Female ups brown.

Flight-period • Mid-April to mid-June in one brood.

♂ S Spain

Habitat • Limestone rocks or dry grassland, often amongst open scrub. LHP: dark-red variety of kidney vetch (*Anthyllis vulneraria*).

Celastrina argiolus Holly Blue

Distribution • Most of Europe, Ireland, England, Wales, & most Mediterranean islands. 0–1900m. Widespread & common.

Description • Male, first & second broods: ups bright sky blue; slightly paler, somewhat silvery veins well defined; very fine, black marginal borders, slightly wider at fw apex. Female, first brood: ups somewhat paler blue; upf black borders very broad at apex & outer margin; uph apex with small black patch; second brood: all ups dark areas extended.

Flight-period • Early April to June & July–August in two broods.

Habitat • Dry or damp bushy places, including woodland margins. LHPs: wide range of shrubs & herbs, including holly (*Ilex aquifolium*), ivy (*Hedera helix*), dogwood (*Cornus sanguinea*), buckthorn (*Rhamnus cartharticus*), & bramble (*Rubus fruticosus*).

Behaviour • Often flies some height above ground. Usually rests with wings closed on leaves in upper parts of LHPs or other shrubs & trees. Does not appear to be greatly attracted to nectar of low plants.

♀ N Belgium

♀ NW Greece

♂ SW France

Glaucopsyche alexis **Green-underside Blue**

Distribution • Much of Europe: Corsica, Sicily, & many Greek islands including Corfu. Absent from British Isles, Atlantic islands, Balearic Islands, Sardinia, & Crete. 25–1500m. Widespread & common.

Description • Uns light grey gc & pattern of spots & unh blue to blue-green basal flush very distinctive – uns without marginal markings (cf. *G. melanops*). Female ups blue basal suffusion variable, sometimes extending to outer margins.

♂ N Greece

Flight-period • April to early July in one brood.

Habitat • Adapted to wide variety of habitat types: dry or damp, grassy & flowery places, often amongst scrub or in woodland clearings. LHPs: several members of pea family, including Spanish broom (*Spartium junceum*), wild liquorice (*Astragalus glycyphyllos*), tufted vetch (*Vicia cracca*), crown vetch (*Coronilla varia*), sainfoin (*Onobrychis viciifolia*), & bladder senna (*Colutea arborescens*).

♀ S Spain

♀ N Greece

♂ (top) & ♀ N Greece

Glaucopsyche melanops Black-eyed Blue

Distribution • N Portugal, Spain, SE France. Italy (W Ligurian Alps). 100–1100 m. Generally widespread but local.

Description • Resembles *G. alexis* but uns with obscure marginal markings & pale grey gc faintly tinged brown. Female ups blue suffusion extremely variable.

Flight-period • Mid-April to May in one brood.

♂ S Spain

Habitat • Open, flowery scrub or light woodland, often with broom. LHPs: several members of pea family, including *Dorycnium decumbens, D. suffruticosum, Lotus hispidus,* & *Anthyllis cytisoides.*

Turanana endymion Odd-spot Blue

Distribution • S Greece (Mt Chelmos & Mt Taygetos). 1500–2300 m. Very sporadic & extremely local.

Description • Male ups dull blue with wide, black marginal borders; unf black pd spots large, with that in s3 conspicuously out of line. Female ups chocolate brown.

Flight-period • Late May to mid-July in one prolonged brood.

Habitat • Open, dry limestone rocks, supporting low-growing, mostly spiny, cushion-forming shrubs, including LHP: *Acantholimon androsaceum.* All known habitats shared with *Lycaena thetis.*

♀ S Greece

Behaviour • Activity largely confined to vicinity of LHP. Both sexes strongly attracted to flowers of thyme (*Thymus*). Males tend to wander considerable distance to drink from damp ground.

Conservation • On Mt Chelmos, human activity poses serious threat to this and other very rare European butterflies.

♀ S Greece

♂ S Greece

Maculinea alcon **Alcon Blue**

Distribution • NE Spain (province of Santander). France, through N Italy & Switzerland to W Denmark, SW Sweden, Lithuania, & C Balkans. 0–1000 m. Extremely sporadic & local throughout range.

Description • Resembles *M. rebeli* closely. Male ups gc dull, powdery blue. Female ups gc greyish brown, often with blue basal suffusion.

Flight-period • Mid-June to mid-August in one brood. Peak emergence normally late July, but depends on locality.

♀ N Belgium

Habitat • Flowery, marshy meadows, usually by rivers or lakes, containing an abundance of LHP: principally marsh gentian (*Gentiana pneumonanthe*).

Conservation • All remaining European colonies, mostly very small, at risk from human interference, especially indirect destruction or damage to habitat from land drainage at well-removed sites.

Maculinea rebeli **Mountain Alcon Blue**

Distribution • N Spain, through C S France, C Italy, & S Poland to S Balkans & N Greece. 600–2250 m. Extremely sporadic & local.

Description • Resembles *M. alcon* closely. Male ups gc brighter blue, lacking violet overtones. Female ups gc greyish brown with more extensive violet-blue basal flush, extending to & accentuating dark discal spots.

Flight-period • Mid-June to July in one brood. Peak emergence normally early July, but much dependent on altitude.

♀ N Greece

Habitat • Dry or damp to wet flowery meadows, often scrub or woodland clearings. LHP: principally cross gentian (*Gentiana cruciata*).

Conservation • Many colonies very small, &, in consequence, especially vulnerable.

Maculinea arion **Large Blue**

Distribution • N & C Spain, to S Fennoscandia, Balkans, & C Greece. Absent from Mediterranean islands except Corsica. 50–2000 m. Sporadic & local in many regions. Indigenous British population extinct.

Description & variation • Male ups bright shiny blue, with narrow greyish borders; pd black markings clearly defined, variable in size. Female similar; ups borders wider; black markings bolder. Both sexes: uns gc light greyish brown;

♂ N Greece

all black markings bold. § 1000–1500 m, *M. a. arion* f. *obscura*. Ups shiny, deep sky blue; black marginal borders wide, pd markings bold, clearly defined. § Above 1500 m. Ups pd areas more grey, further extended, largely obscuring black pd markings; basal flush, dull violet blue, much reduced.

♂ N Greece

Flight-period • Late May to July in one brood.

Habitat • Dry, flowery, grassy places, often amongst scrub or light woodland. LHPs: thyme, including wild thyme (*Thymus serpyllum*) & common wild thyme (*T. praecox*).

Maculinea telejus Scarce Large Blue

Distribution • C W France. E France through N Switzerland & N Italy to S Poland & Hungary. SW Latvia (one known colony). Extinct in Belgium. 700–1600 m. Extremely sporadic & local.

Description • Male ups bright, shiny blue with silvery overtones in outer margin; dark marginal borders wide; dark pd markings clearly defined; uns gc light coffee brown, with black, white-ringed markings well defined. Female ups greyish brown, with darker blue flush extending slightly beyond pd spots; uns darker brown.

Flight-period • Mid-June to mid-August in one brood.

Habitat • Flowery, marshy meadows containing an abundance of LHP: great burnet (*Sanguisorba officinalis*). Habitat very similar to, & often shared with, that of *M. nausithous*.

Behaviour • Often roosts on flower heads of LHP.

Conservation • Most habitats occur in areas of cultivation.

Maculinea nausithous Dusky Large Blue

Distribution • N Spain (Picos de Europa, Soria, & Sierra de la Demanda). E France to C Balkans. 700–1600 m. Extremely sporadic & very local.

Description • Male ups brown, with dusky blue basal flush extending beyond dark pd spots; uns gc café-au-lait, black, white-ringed pd spots well-defined. Female ups brown; uns coffee brown, pd spots well-defined.

Flight-period • Mid-June to mid-August in one brood.

Habitat • Flowery, marshy meadows, containing an abundance of LHP: great burnet (*Sanguisorba officinalis*). Habitat very similar to, & often shared with, that of *M. telejus*.

Behaviour • Often roosts on flower heads of LHP.

Conservation • Many habitats occur in areas of cultivation & human activity.

Iolana iolas **Iolas Blue**

Distribution • S & E Spain, through S France, Switzerland (Rhône Valley), & C & N Italy to Balkans & Greece. 100–1700 m. Very sporadic, extremely local.

Description • Male ups gc pale sky blue; veins lighter, silvery; uns creamy grey; marginal markings, black, white-ringed pd & discal spots well-defined. Female ups greyish brown, with extensive violet-blue basal flush; uns as for male.

♂ C Greece

Flight-period • May to early July in one brood.

Habitat • Dry, bushy places, usually on limestone & in close proximity to LHP: bladder senna (*Colutea arborescens*).

Behaviour • Fast & powerful flight. Males often recorded in areas well-removed from known breeding grounds.

♂ C Greece

♀ NW Greece

Pseudophilotes baton **Baton Blue**

Distribution • N Portugal & N Spain, through France, Corsica, & Sicily, to SW Poland & W Austria. 200–2000 m. Widespread, locally often very common.

Description • Both sexes resemble *P. vicrama*, but smaller. Uns orange submarginal spots generally paler & less conspicuous.

Flight-period • Generally April–June & late July to September in two broods. High altitudes: June–July in one brood.

Habitat • Dry, grassy, & flowery places, often in scrubland or amongst rocks. LHPs: principally thyme, including wild thyme (*Thymus serpyllum*), thyme (*T. vulgaris*), common wild thyme (*T. praecox*); also winter savoury (*Satureja montana*).

Pseudophilotes panoptes **Panoptes Blue**

Distribution • Iberia, south of Cantabrian Mts & Pyrenees. 600–1900 m. Widespread & locally very common.

Description • Resembles *P. baton*. Male unh pale orange or yellow submarginal spots reduced or absent.

Flight-period • Late March to June & July–August in two broods.

♀ S Spain

Habitat • Hot, dry, grassy, rocky, & flowery places. LHPs: principally thyme, including round-headed thyme (*Thymus mastichina*) and *T. villosus*; also winter savoury (*Satureja montana*).

Pseudophilotes vicrama

Distribution • S Finland, through SE Germany & NE Italy to European Turkey & Greece, including Corfu & many Aegean islands. 0–1900 m. Widespread & common.

Description • Resembles *P. baton*, but larger, especially female. Unh orange submarginal spots very well-developed.

♂ N Greece

Flight-period • April to early June & July–August in two broods.

Habitat • Hot, dry, grassy, rocky, & flowery places. LHPs: thyme, including *Thymus longicaulis* & *T. glabresens*; also winter savoury (*Satureja montana*) & *S. thymbra*.

♂ N Greece

♀ NW Greece

Pseudophilotes abencerragus
False Baton Blue

Distribution • SW Iberia. 100–1500 m. Very sporadic; extremely local but often common.

♂ S Spain

Description • Male ups gc steely blue; uns medium grey. Female ups dark brown; uns greyish brown.

Flight-period • April–May in one brood.

Habitat • Hot, dry, flowery places, usually amongst light scrub. LHP: *Cleonia lusitanica.*

♀ S Spain

Pseudophilotes barbagiae Sardinian Blue

Distribution • Known only from Sardinia (Monti del Gennargentu). 800–1500 m. Sporadic, locally common.

Description • Male ups dark greyish brown, blue basal flush shading towards pd area; uns gc greyish brown; unh submarginal orange spots very obscure. Female similar; ups gc uniform dark brown. No similar species occurs in Sardinia.

Flight-period • May–June in one brood.

Habitat • Dry, rocky, flowery places amongst scrub. LHP: thyme (*Thymus herba-barona*).

Pseudophilotes bavius **Bavius Blue**

Distribution • Romania. Republic of Macedonia. NW &
S Greece. European Turkey. 600–1200 m. Extremely sporadic,
very local, & generally very uncommon.

♂ NW Greece

♀ S Greece

Description & variation • § Romania.
Male ups silvery blue. § Republic of
Macedonia. Male ups dull, somewhat greyish
blue; uph submarginal spots absent except
for small, pale orange mark in anal angle. §
Peloponnese. Male ups deep sky blue; uph
orange submarginal spots well-developed.

Flight-period • Mid-May to June in one
brood.

Habitat • Dry, sheltered rocky slopes,
gullies, or small clearings amongst scrub,
usually on limestone. LHPs: *Salvia nutans*,
sage (*S. officinalis*), & *S. verbenaca*.

Conservation • Most known habitats are
very small & appear to be threatened by
over-grazing.

♀ S Greece

Scolitantides orion **Chequered Blue**

Distribution • E Spain, through S France, Corsica, N Italy, & S Switzerland, to Balkans & N Greece. 200–1100 m. Very sporadic & local. S Fennoscandia. 0–300 m. Very local in coastal areas.

Description & variation • Male ups silvery steel blue gc regionally to locally variable in tone & extent. Female gc dark brown to sooty brown with variable blue basal flush, sometimes absent.

Flight-period • Generally May–June in one brood. S Switzerland: April–May & July–August in two broods.

♀ N Greece

Habitat • Dry rocky or sandy places, often with sparse vegetation amongst scrub or scattered trees. LHPs: stonecrops, including white stonecrop (*Sedum album*), orpine (*S. telephium*), & Spanish stonecrop (*S. hispanicum*).

♀ N Greece

♂ N Greece

Chilades trochylus **Grass Jewel**

Distribution • SE Bulgaria. Greece, including Samos, Chios, Rhodes, & Crete. 25–900 m. Colonies widely scattered, often extremely small. Locally abundant.

♂ C Greece

Description • Both sexes, ups gc medium brown.

Flight-period • March–October in several broods.

Habitat • Hot, dry places, often near cultivated ground. LHP: knotgrass (*Andrachne telephioides*).

Behaviour • Both sexes remain very close to LHP.

Conservation • Colonies in cultivated areas are especially vulnerable.

♂ C Greece

Plebejus pylaon **Zephyr Blue**

Distribution • Spain (700–1500 m). S Switzerland & N Italy (1000–2000 m). Balkans & Greece (500–2050 m). In Spain & southern Alps, colonies extremely small & widely dispersed. In SE Europe, far less sporadic, less local, & often abundant.

♀ *P. p. sephirus*, N Greece

Description & variation • *P. p. pylaon* (nominate form) does not occur in Europe. Spain, *P. p. hespericus*: male ups bright turquoise blue, uph black marginal spots sometimes ringed pinkish violet; female ups medium brown, uph submarginal orange spots large. Alps, *P. p. trappi*: male ups darker, violet blue; female ups dark brown, often with blue basal suffusion, uph with dark orange spots in anal angle often absent or replaced by black dots. Balkans & Greece, *P. p. sephirus*: male ups pale blue with violet tone, uph with black dots in anal angle, sometimes with adjacent rust-red marks; female ups medium to light brown, uph submarginal orange

spots usually large, often extending to fw & sometimes forming conspicuous, almost unbroken orange band, less often accompanied by striking, upf orange suffusion. All forms in both sexes: unh orange spots without lustrous, metallic greenish-blue scaling – distinction from *P. argus*, *P. idas*, & *P. argyrognomon*.

Flight-period • Mid-May to early August in one brood; emergence date depends on locality & altitude.

Habitat • Regionally & locally variable. Dry, rocky, flowery places amongst open scrub or sparse woodland; hot, dry, sandy places, often in margins of cultivated ground; grassy places, sometimes in small, damp forest clearings. Distribution of butterfly & its LHPs very closely linked; in Spain, for example, occurrence of LHP appears to guarantee presence of butterfly. LHPs include: (E Europe) *Astragalus exscapus*, *A. dasyanthus*, & *A. parnassi* ssp. *cyllenus*; (Alps) *A. exscapus*; (Spain) *A. turolensis*, *A. alopecuroides*, & *A. clusii*.

Behaviour • In SE Europe, males often assemble in large numbers on damp ground.

Conservation • Many colonies, especially in Spain, appear vulnerable, particularly those close to areas of cultivation. At least two colonies eradicated in Sierra Nevada in last two decades by urban expansion.

♀ *P. p. hespericus*, C Spain

♂ *P. p. sephirus*, N Greece

♂ *P. p. hespericus*, E Spain

♂ *P. p. sephirus*, S Greece

Plebejus argus **Silver-studded Blue**

Distribution • Iberia, through Wales, England, & Corsica to Fennoscandia & Greece, including Corfu, Sicily, & Thassos. 0–2000 m. Widespread, often locally abundant.

Description & variation • Regionally & locally extremely variable. Male ups violet-blue gc variable in tone; ups marginal black borders very variable; uns gc whitish, pale grey, light greyish buff, or pale yellowish brown. Female ups medium brown, with variable blue suffusion, sometimes extending to

♀ S Greece

outer margin of both wings; uph usually with orange spots in anal angle, often large, sometimes extending to fw; uns gc, slightly darker tone; markings averagely bolder. Superficially similar to *P. idas* & *P. argyrognomon* in respect of all characters, but differences – more easily seen than described – are, collectively, generally sufficient for reliable identification.

Flight-period • Single-brooded in northern range, Iberia, Corsica, & at high altitude in southern range (mid-June to late August). Double-brooded elsewhere (May–June & July–August).

Habitat • Most habitat types. Wet or hot, dry, grassy, flowery places usually amongst scrub & rocks, sometimes in light woodland or sheltered, grassy, alpine slopes. LHPs: many genera & species of pea family (Fabaceae), including gorse (*Ulex europaeus*), broom (*Cytisus scoparius*), bladder senna (*Colutea arborescens*); also rockrose (*Helianthemum nummularium*) & ling (*Calluna vulgaris*).

♀ S Greece

♂ S Greece

♂ S Greece

♂ (top) & ♀

♀ N Greece

Plebejus loewii **Loew's Blue**

Distribution · Greece: reported only from Aegean islands of Kos, Patmos, Kalimnos, Tilos, & Rhodes. 0–800 m. Very sporadic & local.

Description · Both sexes resemble *P. pylaon* but not known to occupy same habitats in Europe.

Flight-period · Data limited. Records relate to late May to late June; probably single-brooded.

Habitat · Dry, rocky places containing a spiny vetch – the presumed LHP (very probably *Astragalus* or *Astracantha*).

Plebejus idas Idas Blue

Distribution • Most of Europe, from Spain, Corsica, & Sardinia, to Lappland & northern C Greece. 200–2100 m. Generally widespread but local.

Description & variation • Regionally & locally extremely variable. Male ups violet blue gc variable in tone; ups marginal black borders very variable; uns gc pale grey to light coffee brown. Female ups medium brown, with variable blue suffusion, sometimes extending to outer margin of both wings; uph usually with orange spots in anal angle, often large, sometimes extending to fw; uns gc slightly darker tone; markings averagely bolder. Superficially similar to *P. argus* & *P. argyrognomon* in respect of all characters, but differences –

♀ NW Greece

more easily seen than described – are, collectively, generally sufficient for reliable identification.

Flight-period • Single-brooded (June to early August) or double-brooded (late May to June & July–August) according to altitude & locality.

Habitat • Regionally to locally variable. Closely similar to that of *P. argus*, but includes sheltered sites in Arctic tundra. LHPs: many shrubs & herbs of pea family, including broom (*Cytisus scoparius*), *C. villosus*, hairy greenweed (*Genista pilosa*), *G. depressa*, bird's-foot trefoil (*Lotus corniculatus*), white melliot (*Melilotus alba*), kidney vetch (*Anthyllus vulneraria*); also heather (*Calluna vulgaris*) & sea buckthorn (*Hippophae rhamnoides*).

Behaviour • Adult activity largely confined to proximity of LHPs. Females strongly attracted to nectar-rich plants, such as thyme (*Thymus*). Males sometimes stray to take water from damp ground.

♂ NW Greece

♂ NW Greece

♂ N Norway

Plebejus argyrognomon **Reverdin's Blue**

Distribution • France through NW Switzerland & Italy to S Fennoscandia, Balkans, & N Greece. 200–1500 m. Very sporadic & very local.

♂ N Greece

♀ N Greece

♀ N Greece

Description & variation • Male ups bright blue; black marginal borders fine; uns gc uniform pale grey, somewhat whitish with powdery blue tinge; all markings bright, clearly defined. Female ups brown, sometimes with blue basal flush; submarginal spots on hw well developed, sometimes extending to fw. In northern range female ups blue suffusion more common, often extending to outer margin on both wings. Superficially similar to *P. idas* but, in overall character, especially uns gc & pattern of markings, usually sufficient for reliable identification.

Flight-period • Single-brooded in Scandinavia (late June to late August); double-brooded elsewhere (mid-May to June & late June to July).

Habitat • Dry or damp, grassy, flowery, often bushy places. LHPs: crown vetch (*Coronilla varia*) & wild liquorice (*Astragalus glycyphyllos*).

Vacciniina optilete **Cranberry Blue**

Distribution • Through C Alps (1500–2800 m), & through Germany to Fennoscandia (100–1400 m). Widespread & local. Republic of Macedonia (2000–2200 m). Extremely local.

Description • Female ups dark brown with variable dark violet basal suffusion.

Flight-period • Late June to August in one brood; emergence date in Lappland depends on weather conditions.

Habitat • Dry or damp heathland; moorland; bogs, often in damp forest clearings. LHPs: northern bilberry (*Vaccinium uliginosum*), bilberry (*V. myrtillus*), & cranberry (*V. oxycoccus*).

♂ N Norway

♂ N Norway

Kretania psylorita **Cretan Argus**

Distribution • Crete (Psyloritis Mts & Dikti Mts). 1300–2000 m. Locally abundant.

Description • Resembles *K. eurypilus*, but smaller; uns markings greatly reduced.

Flight-period • June–July in one brood.

Habitat • Open, rocky ground, dominated by spiny cushions of LHP: a species of *Astragalus* or *Astracantha*.

♂ Crete, Greece

Behaviour • Flight very low, & close to LHP.

Conservation • Habitats reportedly threatened by increasing human activity.

Kretania eurypilus **Eastern Brown Argus**

Distribution • Greece: Taygetos Mts & Samos (Mt Kerketefs). 1400–2200 m. Local, but often extremely abundant.

Description • Sexes similar, ups orange submarginal better developed in female.

♀ S Greece

♂ S Greece

Flight-period • June–July in one brood.

Habitat • Sheltered places on calcareous rocks. LHP: *Astracantha rumelica*.

Behaviour • Males may gather in large numbers on wet ground to take moisture.

Eumedonia eumedon **Geranium Argus**

Distribution • From Spain, through S France to Fennoscandia, Balkans, N Sicily, & C Greece. Very sporadic & local in W & SE Europe & S Italy. 0–2400 m; generally below 900 m.

Description & variation • Both sexes: ups dark brown; unh white stripe sometimes much reduced or absent. In cooler climates, generally smaller, uns more grey with reduced markings.

♀ N Greece

Flight-period • Mid-May to mid-August in one brood; emergence date depends mainly on altitude & latitude.

Habitat • Warm, well-sheltered, generally damp, flowery clearings in woodland or scrub. LHPs: cranesbills, including wood cranesbill (*Geranium sylvaticum*), meadow cranesbill (*G. pratense*), & marsh cranesbill (*G. palustre*).

Behaviour • Remains very close to LHP.

♀ N Greece

♂ N Greece

Aricia agestis **Brown Argus**

Distribution • Most of C & S Europe, including Canary Islands, most Mediterranean islands, & S Britain. 0–1900 m. Widespread, generally very common.

Description & variation • First brood unh gc greyish or creamy buff; later brood(s) sandy to rusty brown. In southern range, ups & uns submarginal orange spots better developed, on ups, often forming a continuous band disrupted only by veins – the typical form in Canary Islands & Iberia (including Balearic Islands).

♂ S Greece (1st brood)

Flight-period • Generally April–October in 2–3 broods, according to locality; recorded in all months in Canary Islands.

Habitat • Adapted to wide range of habitat types. Grassy or rocky, flowery places, usually on limestone. LHPs include rockrose (*Helianthemum nummularium*), common storksbill (*Erodium cicutarium*), & bloody cranesbill (*Geranium sanguineum*).

♂ S Spain (2nd brood)

♀ NE Greece (2nd brood)

Aricia artaxerxes **Mountain Argus**

Distribution • Most mountainous areas from Spain through Alps to Balkans & Greece. Scotland. N England. Fennoscandia. Baltic countries. 0–2200 m. Widespread, generally very local.

Description & variation • Regionally & locally extremely variable, superficially resembling some forms of *A. agestis*, but usually easily distinguishable. Scotland, *A. a. artaxerxes*: both

♀ *A. a. montensis*, NW Greece

♂ *A. a. allous*, S Switzerland

♂ *A. a. montensis*, NW Greece

sexes ups gc chocolate brown; upf with very striking white discal spot; ups orange submarginal spots variable, sometimes absent in male, better developed in female, sometimes extending to fw; uns gc fawn; black points in white spots often absent. N England: similar, but upf discal spot more usually black, sometimes faintly ringed white. C Europe & Fennoscandia (0–2200 m), *A. a. allous*: smaller; ups dark brown; fw pointed; ups orange submarginal spots much reduced, usually confined to hw anal angle. S Europe (1000–2200), *A. a. montensis*: larger; fw pointed; ups gc medium brown, orange spots well-developed; uns gc creamy, pale brown, markings bright.

Flight-period • June–September in one brood; emergence date depends on locality, altitude, & seasonal weather conditions.

Habitat • Mostly warm, flowery, grassy, or rocky places, often on limestone or base-rich soils, such as sand-dunes in Fennoscandia. LHPs include rockrose (*Helianthemum nummularium*), bloody cranesbill (*Geranium sanguineum*), wood cranesbill (*G. sylvaticum*), *G. asphodeloides*, *G. cinereum* ssp. *subcaulescens*, & common storksbill (*Erodium cicutarium*).

Aricia morronensis **Spanish Argus**

Distribution • Known only from a few mountains in Spain & France (restricted to Pyrenees – Col du Tourmalet). 900–3000 m. Extremely sporadic, local, but often common.

Description & variation • Size & development of markings extremely variable between isolated populations. Ups gc brown to dark brown; black discal spot sometimes faintly ringed white; upf apex with some white scaling. Sierra Nevada (2050–3000 m), *A. m. ramburi*: smaller; ups gc medium brown; submarginal orange spots absent; uns gc café-au-lait; unh pale yellowy orange submarginal spots inconspicuous, sometimes absent. Soria, *A. m. hesselbarthi* (950–1100 m): larger; ups dark brown; uph orange submarginal spots in anal angle always present; uns gc slightly darker, markings well-developed.

♀ *A. m. hesselbarthi*, Soria, N Spain

♀ *A. m. hesselbarthi*, Soria, N Spain

Flight-period • Single-brooded above 1600 m (July–August); emergence date depends on locality, especially altitude. Double-brooded in Soria (late May to June & July–August).

Habitat • Locally diverse. At highest altitudes, rocky ground with very sparse, low-growing vegetation; at lowest altitudes, grassy, flowery places close to agricultural areas. LHPs: storksbills, including common storksbill (*Erodium cicutarium, E. C. ssp. cicutarium*), *E. ciconium*, soft storksbill (*E. malacoides*), rock storksbill (*E. petraeum* ssp. *crispum*). Butterfly distribution apparently highly correlated with equally rare & local LHPs.

Behaviour • At highest altitudes, flight very fast & low: rests or roosts amongst or under stones. At lowest altitudes, activity is confined largely to vicinity of LHP.

Conservation • Under considerable threat at lowest altitude from agriculture & urbanization. At least some colonies of Sierra Nevada threatened by tourism.

Ultraaricia anteros **Blue Argus**

Distribution • W & S Balkans & Greece. 550–2000 m.
Sporadic & local, but often common.

♀ N Greece

Description & variation • Male gc
variable, bright shiny blue to dull blue; unh
gc greyish or buff in first brood, creamy
white to creamy light brown in later broods.
Female ups uniform medium brown, always
lacking blue basal flush; upf & unh orange
submarginal spots well-developed; unh gc
fawn in first brood, rich, ochreous tan in
later broods. In both sexes, uns colour &
pattern of markings distinctive.

Flight-period • May–September in 1–3
broods according to altitude.

Habitat • Grassy, flowery places, often
amongst scrub or in light woodland. LHPs:
Geranium asphodeloides, bloody cranesbill
(*G. sanguineum*), rock cranesbill
(*G. macrorrhizum*), & ashy cranesbill
(*G. cinereum* ssp. *subcaulescens*).

♀ (2ⁿᵈ brood) N Greece

♂ N Greece

♂ S Greece

Pseudaricia nicias Silvery Argus

Distribution • Spain & France (E Pyrenees, extremely local). Very sporadic & local. SE France, NW Italy, & S & E Switzerland. Sporadic & local. 1000–2300 m. Eastern C Sweden from (60–66°N) & S Finland except W coast. 0–300 m.

Description & variation • Pyrenees & C Alps: male ups pale silvery blue with wide greyish-brown borders; female ups light brown with light brown fringes. Fennoscandia: male slightly larger, ups brighter, paler blue, sometimes with greenish tinge; narrower marginal borders more grey, well-defined.

♀ SE France

Flight-period • S Europe: early July to early September in one brood; emergence date depends on altitude. N Europe: early July to mid-August.

Habitat • Pyrenees & C Alps: warm, sheltered, often damp places with an abundance of flowers & long grasses. Fennoscandia: habitat similar but extends to sheltered coastal districts, including beaches. LHPs: wood cranesbill (*Geranium sylvaticum*) & meadow cranesbill (*G. pratense*).

Albulina orbitulus Alpine Blue

Distribution • C European Alps of France, Italy, S & E Switzerland, & Austria. 1000–2700 m. Mountains of Norway & Sweden 61–64°N. 800–1200 m.

Description & variation • Male ups sky blue with very thin black marginal borders; uns gc greyish buff; unh white spots usually without black points. Female ups brown, often with blue basal suffusion, sometimes extensive.

Flight-period • July–August in C Alps, & June–July in Scandinavia, in one brood.

Habitat • Alpine to subalpine meadows, often in damp situations; sometimes, exposed, steep, flowery slopes with short grass. LHP: alpine milk vetch (*Astragalus alpinus*).

Agriades glandon **Glandon Blue**

Distribution • Spain: Sierra Nevada (2500–3000 m). France (Pyrenees). C Alps (1880–2700 m). Lappland (50–900 m). Extremely sporadic; very local; but often very common.

Description & variation • Pyrenees & Alps, *A. g. glandon*: male ups silvery, somewhat greyish blue gc often very diffuse, especially towards outer margin; upf & uph black discal spots, faintly outlined white; female similar, ups gc greyish brown; both sexes: unh yellow to orange spots variable, rarely absent. Sierra Nevada, *A. g. zullichi*: resembles nominate form; male ups gc slightly darker; female ups gc paler. Lappland, *A. g. aquilo*: smaller; male ups more silvery blue, dark borders narrower, less diffuse; female ups pale, greyish brown; both sexes: unh white marginal spots confluent, generally lacking ocelli.

Flight-period • June–August in one brood, according to altitude, latitude, & seasonal weather conditions.

Habitat • Rocky ground with sparse vegetation or short grass. Colonies often extremely small & often quite isolated, reflecting local distribution of LHP(s): in Spain, vitaliana (*Vitaliana primuliflora*); also, in Alps, blunt-leaved rock-jasmine (*Androsace obtusifolia*) & ciliate rock-jasmine (*A. chamaejasme*); in Lappland, yellow mountain saxifrage (*Saxifraga aizoides*) & purple saxifrage (*S. oppositifolia*).

Behaviour • Flight is fast & low. Uses small stones to shelter from strong winds, under which also roosts. Males sometimes assemble on damp ground to drink.

♂ *A. g. glandon*, SE France

♂ *A. g. zullichi*, S Spain

♀ *A. g. zullichi*, S Spain

♂ *A. g. aquilo*, N Norway

♂ *A. g. aquilo*, N Norway

♀ *A. g. aquilo*, N Norway

Agriades pyrenaicus Gavarnie Blue

Distribution • Cantabrian Mts & Pyrenees. 1550–2200 m.
S Balkans & N Greece. 1500–2300 m. Within each region,
colonies widely dispersed.

Description & variation • Resembles *A. g. glandon*. §
Pyrenees, *A. p. pyrenaicus*. Male ups brighter, silvery blue,
marginal borders narrower; uph submargin whitish. Female
ups greyish brown with bluish tint. § Cantabrian Mts,
A. p. asturiensis. Male ups brighter, submarginal white
markings better developed extending to fw in a continuous

narrow band broken by veins. Female ups
slightly paler brown; uph usually with
obscure submarginal white markings. § S
Balkans & C N Greece, *A. p. dardanus*.
Resembles nominate form; slightly smaller.
Male ups slightly paler. Female without
white marginal marks.

♂ *A. p. pyrenaicus*, SW France

♂ *A. p. pyrenaicus*, SW France

♀ *A. p. pyrenaicus*, SW France

Flight-period • June–July in one brood.

Habitat • Open, rocky slopes or dry, grassy places with short turf. LHP: rock-jasmine (*Androsace villosa*).

Behaviour • Both sexes appear to remain close to area of LHP.

♀ *A. p. dardanus*, N Greece

Cyaniris semiargus **Mazarine Blue**

Distribution • N Portugal, N & E Spain (sporadic & local), through France to N Fennoscandia (rare & local N of Arctic Circle), to Balkans & Greece. Absent from Mediterranean islands except Sicily. Extinct in Britain. 0–2200 m. Locally common.

Description & variation • N Europe: female ups invariably blue. S Spain: female ups basal areas sometimes extensively blue. NW Greece: female unh often with brown, dark grey, or orange spot in anal angle.

§ C Greece, nominate form replaced by f. *parnassia*. Small. Male ups brighter blue; narrow, black marginal borders better defined; unh rarely with orange spot in anal angle. Female uph occasionally with one or two orange spots in anal angle. § S Greece

♂ *C. s. semiargus*, NW Greece

♂ *C. s. semiargus*, NW Greece

♀ *C. s. semiargus*, NW Greece

♂ *C. s. helena*, S Greece

(650–1800 m), f. *helena*. Resembles f. *parnassia* but male uns submarginal orange markings much better developed. Female ups invariably brown; ups & uns orange markings much better developed.

Flight-period • Most of Europe: May–October in two or three broods according to climatic conditions. S Greece: late April to early July in one brood.

Habitat • Grassy, flowery, often damp places; meadows; hayfields; scrub or woodland clearings. LHPs: in most of Europe, red clover (*Trifolium pratense*); in S Greece, *T. physodes*.

Behaviour • Males often assemble on damp patches to drink. In S Greece, in calm, warm conditions, adults often assemble in large numbers on grass stems to bask in late afternoon sun.

♀ *C. s. helena*, S Greece

♀ *C. s. helena*, S Greece

Agrodiaetus iphigenia **Chelmos Blue**

Distribution • S Greece (Mt Chelmos & environs). 1100–1750 m. Very scarce & local.

Description • Female readily distinguishable from other *Agrodiaetus* species by dark, chocolate brown ups & distinctive white fringes. No other 'blue' *Agrodiaetus* species occurs in same habitat.

Flight-period • Late June to July in one brood.

♂ S Greece

Habitat • Mostly dry, open ground above tree-line; sometimes small woodland clearings. LHP: white sainfoin (*Onobrychis alba*).

Behaviour • Males occasionally visit damp ground to drink, invariably feeding with closed wings. Females rarely leave confines of breeding ground. Both sexes bask with partially open wings early morning & late afternoon.

Conservation • Intensive grazing of habitat, coupled with extensive use of host mountain for recreational purposes, appears to pose significant threat. Much of larval host-plant population appears to owe its survival to protective, browse-deterring, spiny plants amongst which it is usually found.

♂ S Greece

♀ S Greece

♀ S Greece

Agrodiaetus damon **Damon Blue**

Distribution • N & E Spain, through S France, N & C Italy (Apennines) to S Poland, Estonia, Latvia (extremely rare), Balkans, & NW Greece. 1000–2100 m. Sporadic, locally common.

♂ C Greece

Description & variation • Baltic countries: larger; male ups deeper blue; wing veins conspicuous.

Flight-period • Mid-July to August in one brood.

Habitat • Dry scrub, open coniferous woodland, or sheltered gullies on open grassy slopes. LHPs include mountain sainfoin (*Onobrychis montana*) & white mountain sainfoin (*O. alba*).

Behaviour • Males sometimes visit damp ground to drink.

♂ E Spain

♀ C Greece

♀ C Greece

Agrodiaetus dolus **Furry Blue**

Distribution • N & E Spain, through S France to NW & S Italy. 500–1500 m. Sporadic & very local.

Description & variation • Male ups pale, silvery blue; upf brown basal androconial patch with slightly roughened appearance. Both sexes: unh white stripe absent or vestigial. § S France (Aveyron, Herault, Gard, & Lozère; 500–1000 m), f. *vittatus*. Male ups gc whitish or pale bluish grey, basal area with bluish suffusion; brownish veins well defined. Female ups gc darker than nominate form. § Peninsular Italy (600–1100 m), f. *virgilius*. Male ups gc white; marginal border, narrow, internally somewhat diffused, especially along veins; upf with pale blue basal suffusion; brown androconial patch rough. Female ups darker brown than nominate form.

♂ S France

Flight-period • Mid-July to August in one brood.

Habitat • Flowery, grassy places amongst scrub; untended margins of cultivated ground. LHP: sainfoin (*Onobrychis viciifolia*).

Agrodiaetus ainsae **Forster's Furry Blue**

Distribution • N Spain: Burgos, Alava, & Huesca. 950–1200 m. Locally very common.

Description • Male ups pale blue; fw androconial patch conspicuous, uns markings closely similar to other members of the genus. Female ups brown, otherwise markings similar. No other 'blue' *Agrodiaetus* species occurs in same habitat.

Flight-period • July–August in one brood.

Habitat • Dry, grassy, bushy places or clearings in light woodland. LHP: sainfoin (*Onobrychis viciifolia*).

Agrodiaetus escheri **Escher's Blue**

Distribution • Spain: Sierra Nevada to Cantabrian Mts & Pyrenees. S France, through S Switzerland & Italy to Balkans & Greece. Sporadic, locally common. 300–2000 m.

Description & variation • Male ups black borders very narrow; uph marginal border vaguely undulate. Female ups submarginal orange spots variable, ranging from 2–3 diffuse spots in anal angle hw, to a continuous band hw & fw. Both

♀ E Spain

sexes: unf without cell-spot; all markings bold. § Ligurian Alps. Smaller. Male ups paler blue; black borders slightly wider; uns gc tending to white; uns markings slightly reduced. Female ups submarginal spots well-developed. § Balkans & Greece. Male ups silvery blue, sometimes with greenish reflections; marginal black borders 1–2 mm. Female ups resembles nominate form. Both sexes: uns markings prominent.

Flight-period • Mid-May to August in one brood. Emergence date depends on locality & altitude.

Habitat • Flowery, often dry rocky places, usually sheltered by bushes; sometimes damp woodland clearings. LHPs: milk-vetches, principally Montpellier milk-vetch (*Astragalus monspessulanus*).

♀ E Spain

♂ N Greece

♂ N Greece

Agrodiaetus amanda **Amanda's Blue**

Distribution • N, S & E Spain, through S France, Italy, C Switzerland, & E Germany to S Fennoscandia, Balkans, Greece, & European Turkey. Absent from Mediterranean islands except NE Sicily & Lesbos. 100–2000 m.

Description & variation • Male ups shining blue; upf margins shaded greyish brown. § Scandinavia. Female ups often extensively suffused blue.

Flight-period • Late May to July in one brood.

♀ S Greece

Habitat • Warm, grassy, often damp places containing an abundance of LHP, usually sheltered by scrub or light woodland. LHPs: vetches of genus *Vicia*, including *V. cracca* & *V. villosa*.

♀ S Greece

♂ S Greece

Agrodiaetus thersites **Chapman's Blue**

Distribution • S Portugal. Spain. France (Pyrenees) to 51°N in Germany. E & SE Europe. Absent from Mediterranean islands except Sicily, Samos, Kos, & Rhodes. European Turkey. 0–1500 m.

♂ W Switzerland (2nd brood)

♂ S Greece (1st brood)

Description • Male upf androconial patch conspicuous. Spring brood: both sexes unh gc grey; female ups blue suffusion often extending to outer margins. Summer brood: both sexes unh sandy brown; female ups brown; unf yellowish grey. Both sexes: unf without cell-spot (cf. *Polyommatus icarus*).

Flight-period • Generally April–June & June–August in two broods. Emergence of summer brood may be delayed or precluded by dehydration of LHPs.

Habitat • Warm, dry, rocky, bushy places; grassy clearings in scrubland; meadows; areas of neglected cultivation. LHPs: sainfoin (*Onobrychis viciifolia*) & cockscomb sainfoin (*O. caput-galli*).

♀ S Spain (1st brood)

The following eight species are characterized by brown uppersides in both sexes, hence the name of the group – the 'anomalous blues'. Although distinction is difficult in some cases, the presence or absence of a white stripe on v4 unh coupled with distributional data is usually sufficient for species identification. A rough, somewhat hairy androconial patch in the basal and discal area of male upf and a lighter tone of brown in female ups gc allows easy differentiation of the sexes.

Agrodiaetus admetus Anomalous Blue

Distribution • Hungary, through S Balkans to Greece & European Turkey. 50–1500 m. Sporadic, especially in northern range; locally common.

Description • Male unh white stripe absent or poorly developed. Female uns marginal & submarginal markings prominent; white chevron along v4 often reduced or absent.

♀ S Greece

Flight-period • Mid-June to July in one brood.

Habitat • Hot, dry, grassy, flowery scrubland. LHPs: sainfoin (*Onobrychis viciifolia*) & cockscomb sainfoin (*O. caput-galli*).

♂ S Greece

♂ S Greece

Agrodiaetus fabressei
Oberthur's Anomalous Blue

Distribution • Spain: provinces of Malaga, Granada, Jaen, Albacete, Teruel, Cuenca, Soria, & Burgos. Locally abundant. 900–1750 m.

♂ Spain

Description • Resembles *A. admetus*. Male unh without white stripe along v4. Female unh small white chevron sometimes present on v4. Both sexes: uns markings much reduced or absent.

Flight-period • Late June to August in one brood.

Habitat • Rocky gullies with scrub; dry, grassy, & flowery places, including roadside verges. LHPs: sainfoins, including *Onobrychis viciifolia* & *O. peduncularis*.

♀ Spain

Agrodiaetus agenjoi
Agenjo's Anomalous Blue

Distribution • NE Spain: known only from Gerona, Barcelona, & Lérida. 700–1500 m.

Description • Resembles *A. fabressei* very closely. Male ups dark brown; unh without white stripe along v4.

Flight-period • July–August in one brood.

Habitat • Rocky gullies with scrub; dry grassy slopes. LHP: sainfoin (*Onobrychis viciifolia*).

Agrodiaetus humedasae
Piedmont Anomalous Blue

Distribution • NW Italy: known only from Valle d'Aosta (Cogne Valley). 800–950 m.

Description • Male uns gc pale creamy brown; unh without white stripe. No other member of this genus occurs in same habitat.

Flight-period • Mid-July to August in one brood.

Habitat • Flowery slopes amongst scrub & small trees. LHP: sainfoin (*Onobrychis viciifolia*).

Conservation • Known habitat very restricted & in close proximity to human habitation & areas of cultivation.

♂ NW Italy

♂ NW Italy

Agrodiaetus aroaniensis
Grecian Anomalous Blue

Distribution • Greece. Widespread but generally very local. 800–1550 m.

Description • Uns gc distinctive, uniform yellowish grey; unh lacking white stripe on v4 (cf. *A. ripartii*).

Flight-period • July to early August in one brood.

Habitat • Dry, bushy, or rocky places; sometimes in sparse woodland. LHP: *Onobrychis arenaria*.

Agrodiaetus ripartii
Ripart's Anomalous Blue

Distribution • NE & E Spain, SE France, NW Italy, S Poland, S Balkans, Greece, & European Turkey. 50–1800 m. Sporadic & local in some regions.

♂ N Greece

Description & variation • Unh white stripe along v4 prominent (cf. *A. aroanensis* & *A. fabressei*). § Greece. Unh white stripe often narrower & fainter.

Flight-period • Late June to early August in one brood.

Habitat • Dry, bushy places, often grassy; sometimes in light woodland. LHPs: sainfoins (*Onobrychis* species).

♂ S Greece

♀ S Greece

Agrodiaetus nephohiptamenos
Higgins' Anomalous Blue

Distribution • N Greece: Mt Pangeon, Mt Phalakron, & Mt Orvilos. Bulgaria: Mt Alibotush. 1500–2000 m. Usually very uncommon.

Description • Male hw fringes conspicuously white. No other member of this genus occurs in same habitat.

♂ N Greece

Flight-period • Mid-July to late August in one brood. Emergence date very dependent upon local weather conditions.

Habitat • Open grassy, flowery slopes above tree-line. LHP: mountain sainfoin (*Onobrychis montana* ssp. *scardica*).

♀ N Greece

Agrodiaetus galloi **Gallo's Anomalous Blue**

Distribution • Known only from mountains of S Italy, including Mte Pollino. 1100–2200 m, most abundant at 1750–1900 m.

Description • Ups chestnut brown with olive tones; fringes creamy white; unh white stripe on v4 conspicuous. Male upf sex-brand conspicuous.

Flight-period • July–August in one brood.

Habitat • Grassy slopes, beechwood clearings, & screes. LHP: probably sainfoin (*Onobrychis* species).

Neolysandra coelestina **Pontic Blue**

Distribution • S Greece: Mt Chelmos & environs. 700–1800 m.

Description & variation • Male upf wide, dark marginal borders well defined (cf. *C. semiargus*). Female ups brown, usually with obscure orange spots in anal angle. Male unh sometimes with pale orange spots in s1–3.

Flight-period • Late May to mid-June in one brood.

Habitat • Sheltered, grassy gullies & hollows; woodland clearings; screes. LHP: tufted vetch (*Vicia cracca* ssp. *stenophylla*).

Behaviour • Rarely strays far from LHP. Males sometimes visit damp patches to drink.

Conservation • At potential risk from overgrazing in some localities.

♀ S Greece   ♂ S Greece

Plebicula dorylas Turquoise Blue

Distribution • N & E Spain, through C & S France to
S Sweden, Lithuania (very rare & local), Balkans, & Greece.
75–2300 m. Generally sporadic & local.

Description • In Spain, female ups sometimes with blue
basal flush; wing bases always with a few blue scales (cf. female
P. nivescens).

♂ N Greece

Flight-period • Low altitudes: May–June
& July–August in two broods. High altitudes
(generally above 1500 m): late June to August
in one brood.

Habitat • Grassy, flowery places, usually
amongst scrub at lower altitudes; sheltered
hollows on open, grassy slopes on high
mountains. LHP: kidney vetch (*Anthyllis
vulneraria*).

Behaviour • Females are not commonly
observed (cf. *P. nivescens* & *P. golgus*).

♀ N Greece

♀ N Greece

Plebicula golgus **Nevada Blue**

Distribution • S Spain: known only from Sierra Nevada (2400–3000 m) & Sierra de la Sagra (1900–2350 m). Locally abundant.

Description & variation • Both sexes: uns resembles *P. dorylas* but gc darker. § Sierra de la Sagra, f. *sagratrox*. Both sexes readily distinguishable from nominate form by overall brighter appearance.

♂ *P. g. golgus* S Spain

Flight-period • Late June to late July in one brood.

Habitat • Exposed, rocky slopes with low-growing, mostly very sparse, vegetation. LHP: kidney vetch (*Anthyllis vulneraria* ssp. *arundana*).

Behaviour • At peak emergence, both sexes observed in roughly equal abundance (cf. *P. dorylas* & *P. nivescens*).

P. g. golgus ♀ S Spain

P. g. sagratrox ♀ S Spain

Plebicula nivescens **Mother-of-pearl Blue**

Distribution • SW Spain to S Cantabrian Mts & S Pyrenees. 1000–1900 m. Widespread, generally very local.

Description • Female ups without blue basal flush – distinction from *P. dorylas* with which it sometimes occurs.

♂ E Spain

Flight-period • Late May to early August in one prolonged brood.

Habitat • Hot, dry, limestone rocks, usually amongst scrub. LHP: kidney vetch (*Anthyllis vulneraria*).

Behaviour • Females appear to be secretive & not commonly observed (cf. *P. dorylas* & *P. golgus*).

♂ E Spain

♀ E Spain

Meleageria daphnis **Meleager's Blue**

Distribution • From E Spain (very local & uncommon), through S France & Italy to C & SE Europe. Very rare & local in S Germany & S Poland. Widespread, locally common in S Balkans & Greece. Absent from Mediterranean islands except Sicily & Aegean island of Simi. 200–1700 m.

Description & variation • Hw scalloped, more strongly so in female. Female ups gc sometimes greyish brown; markings outlined by whitish or greyish suffusion, sometimes extending along veins; the dominant form in some localities.

Flight-period • Mid-June to August in one brood.

Habitat • Grassy or bushy, generally dry places. LHP: crown vetch (*Coronilla varia*).

♀ N Greece (colour variant)

♀ N Greece (colour variant)

♂ N Greece

Lysandra coridon **Chalk-hill Blue**

Distribution • N & E Spain through most of Europe, including S England, Corsica, & Sardinia, to Lithuania & Greece. 100–2200 m. Somewhat local, but widespread & often abundant.

Description • Male ups pale silvery blue with faint yellowish reflections. Female ups usually brown, sometimes with blue basal suffusion, rarely extending to submargin.

Variation

Subject to marked local to regional variation, especially in Spain. § Montes Universales (1050–1800 m), *L. c. caelestissima.* Male ups gleaming sky blue. Female ups usually brown, very rarely silvery sky blue. § N Spain (600–1950 m), *L. c. asturiensis.* Male ups shiny, silvery blue – blue shade locally to regionally variable. In some localities, female ups colour as for male in most specimens, but proportion of 'normal' females, i.e., with brown ups, to blue form is regionally extremely variable. Hybrids with *L. bellargus* have been reported from most populations.

Flight-period • Late June to early October in one brood. Emergence date depends on locality & altitude.

Habitat • Dry, flowery, places often with short grass; usually on limestone, but always on alkaline soils, e.g., chalk downs of S England. In S Europe, habitats characteristically rockier & often associated with open pinewoods. LHP: horse-shoe vetch (*Hippocrepis comosa*).

Behaviour • Males often gather in large numbers to take

water from damp ground. Both sexes roost, usually in tight communities, at tops of grass stems, invariably with head pointing down, evidently to eliminate risk of wings being blown open & damaged in event of wind bending grass stem.

Conservation • Loss of colonies in some regions, e.g. England, due largely to interference with, or loss of, habitat.

♂ N Greece

♀ N Greece

L. c. caelestissima ♂ E Spain

L. c. caelestissima ♀ E Spain

L. c. caelestissima ♀ E Spain

Possible hybrid between *L. coridon* & *L. bellargus*; ♂ S Greece

Lysandra philippi
Macedonian Chalk-hill Blue

Distribution • N Greece: known only from Mt Pangeon & Mt Phalakron. 600–1900 m. Local & uncommon.

Description & variation • Resembles *L. coridon* closely, but no other members of group known to occur within same habitat. On Mt Pangeon, female ups with extensive blue suffusion in about 50% of specimens.

Flight-period • Early July to August in one brood.

Habitat • Dry, open scrubland & grassy slopes. LHP: horse-shoe vetch (*Hippocrepis comosa*).

♀ N Greece

♀ N Greece

♂ N Greece

Lysandra hispana **Provençal Chalk-hill Blue**

Distribution • E Spain, through S France to NW Italy.
400–1000 m. Locally common.

Description • Male ups somewhat dull, bluish grey with
silvery-yellow reflections; unh somewhat darker & greyer than
L. coridon, with which it flies in most localities in second
brood. Female indistinguishable from *L. coridon*.

Flight-period • Mid-April to late June & August to early
October in two broods.

Habitat • Dry, flowery, grassy places, often amongst scrub.
LHP: horse-shoe vetch (*Hippocrepis comosa*).

♂ NW Italy

♂ NW Italy

♀ NW Italy

♀ NW Italy

Lysandra albicans **Spanish Chalk-hill Blue**

Distribution • S, C, & E Spain. 500–1500 m. Locally
abundant.

Description & variation • § S Spain. Large; male ups very
pale, almost white. § N & E Spain. Smaller; male ups with
bluish-grey tint. Generally, a variable species.

♂ SW Spain

Flight-period • Mid-June to August in
one brood.

Habitat • Dry rocky places, often with
sparse vegetation. LHPs: principally horse-
shoe vetch (*Hippocrepis comosa*).

♀ S Spain

Lysandra bellargus **Adonis Blue**

Distribution • Most of Europe from Mediterranean coast to S England & Lithuania. Absent from Mediterranean islands except Mallorca & C Sicily. 100–2000 m.

Description & variation • Uns resembles *L. coridon*; unh submarginal orange spots usually better developed. Females ups sometimes with blue suffusion, especially in C Spain.

♂ S Spain

Flight-period • Generally, mid-May to June & late July to mid-September in two broods. Possibly one brood in S Greece.

Habitat • Generally dry, grassy places, often amongst scrub. LHPs: horse-shoe vetch (*Hippocrepis comosa*), less commonly crown vetch (*Coronilla varia*).

Conservation • Becoming increasingly rare in England.

♀ S Spain

♀ N Greece

Polyommatus icarus **Common Blue**

Distribution • Widespread & very common in most of
Europe, including all major & most smaller Mediterranean
islands. Reported from Fuerteventura, Lanzarote, & Tenerife
(Canary Islands), but absent from Madeira & Azores.
0–2900 m.

♀ NW Greece

♀ S Spain

Description & variation • Male upf
without androconial patch. Female ups gc
brown, often with blue basal & discal
shading; ups blue suffusion sometimes
extensive, a prevalent form in some localities,
e.g. N & W Ireland & NW Scotland. Both
sexes: unf with black cell-spot (cf.
Agrodiaetus thersites). Subject to appreciable
regional variation, especially size.

Flight-period • N Britain, Lappland, &
high altitudes in S Europe: June–July in one
brood. Elsewhere: late March to early
November in 2+ broods, depending on
regional & local climatic conditions.

Habitat • Almost all habitat types. One of
Europe's most adaptable butterflies. Occurs
close to inhospitable summit of Sierra
Nevada, as well as equally demanding hot,
dry shoreline of nearby Spanish coast. LHPs
include many members of pea family,
commonly bird's-foot trefoil (*Lotus
corniculatus*).

♂ S Spain

♂ NW Greece

Polyommatus andronicus **Phalakron Blue**

Distribution • SW Bulgaria (Mt Alibutush. 1400 m).
N Greece (Mt Orvilos, Menikion Mts, & Phalakron Mts.
1000–1800 m.) Locally common.

Description • Resembles *P. icarus* very closely. Both sexes
larger. Male ups gc slightly darker & shinier. Female ups largely
without blue basal scales. Main distinction based on genitalia
but unh ultraviolet reflectance pattern also differs from that of
P. icarus, thus reducing risk of confusion in courtship.

Flight-period • Late June to early July in one brood. Flies
with second brood of *P. icarus*.

Habitat • Sheltered grassy or rocky slopes. LHP(s) not
reported.

Polyommatus eroides **False Eros Blue**

Distribution • C E Europe to NW Greece. 950–2100 m.
Sporadic & very local but often very common.

♂ NW Greece

Description • Male ups black marginal
borders 1–2 mm wide; uns gc uniform dove
grey. Female ups gc uniform medium brown,
sometimes with very faint greyish tint,
lacking blue basal suffusion (cf. *P. icarus*).
Both sexes: uns markings well-developed.

Flight-period • Mid-June to late July in
one brood.

Habitat • Open, dry, or damp flowery
banks or slopes with short grass, often
amongst rocks above tree-line. LHP: *Genista
depressa*.

Behaviour • Males often gather in large
numbers on damp ground.

♀ NW Greece

♀ NW Greece

Polyommatus eros **Eros Blue**

Distribution • Sporadic & local on mountains of S Europe from E Spain (Pyrenees), through C Alps & Apennines to Macedonia. 1200–2700 m, generally above 1800 m.

Description • Male ups gleaming sky blue, with conspicuous, narrow black borders contrasting with white fringes. Female ups greyish brown; wing-bases usually dusted with pale blue scales; orange submarginal spots variable in size & colour.

♀ SE France

Flight-period • July–September in one brood. Emergence date depends on altitude.

Habitat • Grassy, flowery slopes, usually with short turf. LHPs: silky milk-vetch (*Oxytropis halleri*) & yellow milk-vetch (*O. campestris*).

Behaviour • Males often gather in large numbers on damp ground.

♀ SE France

Polyommatus menelaos **Taygetos Blue**

Distribution • Greece: known only from Taygetos Mts. 1250–2000 m.

Description • Resembles *P. eros*. Both sexes: larger, uns gc paler, tending to white.

♂ S Greece

Flight-period • Early June to late July in one brood.

Habitat • Sheltered gullies & hollows, generally above tree-line. LHP: *Astragalus taygeteus* (endemic to Taygetos Mts).

Behaviour • In hot conditions, males gather in large numbers on damp ground. Often found with *Kretania eurypilus*.

♂ S Greece

♀ S Greece

Riodinidae

This large family, whose worldwide distribution centres on tropical America, is represented in Europe by a single species, which, although resembling a 'Fritillary', is more closely related to the Lycaenidae.

Hamearis lucina
Duke of Burgundy Fritillary

Distribution · Spain, through most of Europe including England (very local) to SE Sweden, Latvia, & C Greece. Absent from Mediterranean islands except N Sicily. 50–1600 m. Widespread, locally common.

♀ S France

Description & variation · Female ups orange markings better developed, especially on fw. § S Europe. Usually larger. Second brood: ups dark markings more extensive, sometimes obscuring orange uph gc.

Flight-period · May–June in one brood, or April–June & July–September in two broods, depending on altitude & latitude.

Habitat · Grassy, flowery places in woodland clearings or margins. LHPs: commonly primrose (*Primula vulgaris*).

Behaviour · Often sits on leaves of LHP with wings half-open.

♂ S France

Libytheidae

Closely related to the Nymphalidae. Represented in Europe by a single species. The conspicuous palpi are over three times the length of the head.

Libythea celtis Nettle-tree Butterfly

Distribution • Portugal & Spain through S Europe to Balkans & Greece. Reported from Corsica, Elba, Sardinia, Sicily, Lipari Islands, Malta, Crete, & Lesbos. 400–1500 m. Very sporadic. Often locally common in Greece, but generally scarce in western and northern range.

♂ S Greece

Description • Male unh greyish. Female unh light brown, with buff or pinkish tints in fresh specimens.

Flight-period • June–August in one brood, reappearing, after hibernation, March–April.

Habitat • Open, bushy areas or light woodland. LHP: nettle tree (*Celtis australis*).

Behaviour • Disperses over wide area in mid- to late summer. Vagrant specimens recorded at 2300 m. Freshly emerged adults often assemble in large numbers to drink on damp ground. Hibernation appears to be preceded, without disruption, by aestivation in late July to August.

♀ S Greece

Danaidae

Of this large family of large butterflies, confined mostly to the tropical region, only two have been recorded in Europe. Both are well-known migrants, &, indeed, the Milkweed butterfly has achieved fame for its extraordinary ability to cross the Atlantic Ocean. When settled in its breeding grounds, however, its somewhat weak, 'flapping' flight rather belies this exceptional proficiency.

Danaus plexippus **Milkweed, Monarch**

Distribution • Canary Islands & S Spain (province of Malaga): resident in coastal districts (0–100 m), occasional as vagrants/migrants to 400 m. Occasional as migrants in Azores, Lanzarote, S & W Portugal, Gibraltar; more rarely, SW Ireland, SW England, & S France. Very rarely recorded in Scotland: few specimens have been noted since 1880, of which one was captured in Shetland Isles in 1941, another in Sutherland in 1946. 0–400 m, generally below 200 m. Resident colonies extremely local, especially in Spain.

Description • Male uph with small sex-brand on v2. Both sexes: ups gc dusky orange; wide, black borders with double row of pale spots; veins conspicuously lined black.

Flight-period • Continuously brooded. Recorded in all months in places of residence.

Habitat • Hot, dry coastal gullies, usually close to cultivated ground & urban areas. LHPs: milkweed (*Asclepias curassavica*) & *Gomphocarpus fruticosus*.

Behaviour • A well-known migrant from C & N America.

Conservation • In S Spain, use of herbicides & insecticides, weed-burning, & rubbish-disposal threaten extinction of resident colonies, which are mostly small & isolated.

♂ (right) & ♀ S Spain

Danaus chrysippus **Plain Tiger**

Distribution • Canary Islands: resident in coastal districts on La Palma, Gomera, & Fuerteventura (0–600 m). Recorded from mainly coastal districts of Tenerife, Gran Canaria, S Spain, S France, Corsica, Sardinia, Sicily, W Italy, Montenegro, Albania, Corfu, W & S Greece, & Crete, mostly as migrants, but breeding populations appear to have become established in some localities.

Description • Male uph with black sex-brand on v2. Both sexes: ups gc dusky orange; white spots in dark apical area conspicuous; uph with three black discoidal spots.

Variation • Uph gc sometimes partly or wholly white.

Flight-period • Canary Islands: continuously brooded; recorded in all months. European Mediterranean region: records span May–October.

♂ S Turkey

Habitat • Hot, bushy, & rocky places, often coastal gullies close to cultivated areas & human habitation. LHPs include milkweed (*Asclepias curassavica*), *Gomphocarpus fruticosus*, & *Caralluma burchardii*.

Behaviour • Powerful, wide-ranging African migrant.

Conservation • As in case of *D. plexippus*, many colonies threatened by human activity.

Nymphalidae

This large family contains several of the most colourful & better known European butterflies. A few of these, notably the Peacock, Small Tortoiseshell, Painted Lady, and Red Admiral, are regular summer visitors to gardens, parks, and even town centres in their quest for nectar-rich plants. A few are migrants. The Painted Lady is perhaps the best known, and in favourable years is able to extend its range from North Africa to well within the Arctic Circle. The largest group, the Fritillaries, occur throughout the region, with some species confined to within the Arctic Circle. Many smaller members of this group are fond of resting in sunshine with open wings. Characteristically, flight is graceful & gliding.

Charaxes jasius **Two-tailed Pasha**

Distribution • Mainly Mediterranean coastal districts, including Balearic Islands, Corsica, Sardinia, Sicily, Corfu, Crete, Samos, Ikaria, Chios, & Rhodes; also, W Portugal & inland regions of S Spain & S France. 0–1200 m. Sporadic, locally common.

Description • Ups brown, with bold orange, marginal markings; hw with very distinctive twin 'tails'.

Flight-period • May–June & August to mid-October in two broods.

Habitat • Hot, dry places usually with dense scrub containing an abundance of LHP: strawberry tree (*Arbutus unedo*).

Behaviour • Flight very fast & powerful. Strongly attracted to fermenting fruit.

♂ N Greece

Apatura iris **Purple Emperor**

Distribution • N Portugal & N Spain through France, S England, & N Italy to Denmark, S Sweden (restricted to Skane), Estonia, & NW Greece. 50–1500 m. Generally sporadic, sometimes locally common.

Description • Male ups gc dark brown with strong purple iridescence when viewed at oblique angle. Female ups brown, lacking purple reflections. Both sexes: upf dark spot in s2 obscured by gc (cf. *A. ilia*).

Flight-period • Mid-June to mid-August in one brood.

Habitat • Margins or clearings of mature deciduous woodland bordered by LHPs: willows & sallows, principally goat willow (*Salix caprea*). *A. iris* & *A. ilia* often share same habitat & sometimes same LHP.

♂ S Switzerland

♂ NW Greece

♀ S Switzerland

Behaviour • Males strongly attracted to carnivore excrement, carrion, human perspiration, hot road-tar, & fumes of petroleum spirit. Both sexes often rest on leaves of trees (commonly oak) several metres above ground.

Conservation • Physical structure of habitats, especially size & position of LHP, appears to be quite critical to long-term viability of colonies, habitat accordingly sensitive to interference.

Apatura ilia **Lesser Purple Emperor**

Distribution • N Iberia, through France & NW Italy to Latvia & N C Greece. 300–1300 m. Very sporadic & local.

Description & variation • Male ups dark brown with purplish reflections at oblique angles; upf dark spot in s2 well-defined, ringed orange (cf. *A. iris*). Female ups dark brown without purplish reflections. § Both sexes, f. *clytie*.

♂ NW Greece

Ups gc sometimes replaced by yellow to orangey brown; ups white markings, except apical spots, replaced by yellow. Resembles *A. metis*, but uph yellow pd band clearly disrupted at v4 (habitats rarely shared).

Flight-period • Late May to July in one brood, or June & August–September in two broods.

Habitat • Deciduous woodland clearings, bordered by LHP, often wooded river banks supporting LHPs in abundance. LHPs include aspen (*Populus tremulae*), white poplar (*P. alba*), black poplar (*P. nigra*), & white willow (*Salix alba*).

Behaviour • As for *A. iris*.

♂ NW Greece

♀ NW Greece

Apatura metis **Freyer's Purple Emperor**

Distribution • Associated largely with river systems in
E Austria, Hungary, Slovenia, N Serbia, & Bulgaria, to
N Greece & European Turkey. 0–650 m. Sporadic & usually
extremely local at limits of range.

♂ NE Greece

♀ NE Greece

Description & variation • Resembles
A. ilia f. *clytie*. Generally smaller; ups
brighter; uph dark pd spots obscured by
dark background; uph outer edge of pale
discal band well-marked by discontinuity at
v4. Usually they do not share same habitats.

Flight-period • Late May to June & mid-
July to August in two broods.

Habitat • Wooded river margins
containing mature specimens of LHPs:
principally white willow (*Salix alba*).

Behaviour • In very hot conditions, adults
rest for long periods on leaves in higher
branches of LHPs. Males strongly attracted
to carnivore excrement or carrion, & often
take water from damp ground. Both sexes
attracted to sap of wounded tree-bark.

♀ NE Greece

Limenitis populi **Poplar Admiral**

Distribution • N & E France (very local in Brittany) through N Italy to Fennoscandia, Bulgaria, & C N Greece. Extinct in Denmark. 200–1500 m. Very sporadic & local, especially near limits of range.

♂ N Greece

♂ N Greece

Description • Female ups markings better developed.

Flight-period • Late May to late July. Emergence date seasonally variable in many localities.

Habitat • Clearings & margins of mature, deciduous woodland bordered by LHP: principally aspen (*Populus tremula*).

Behaviour • Males attracted to carnivore excrement, human perspiration, hot road-tar, petroleum spirit, & damp wood-ash.

Limenitis reducta **Southern White Admiral**

Distribution • N Portugal, SE, & N Spain, through S France to C Balkans & Mediterranean region, including most larger islands except Balearics & Crete. 0–1650 m. Also, NW France (Brittany & Normandy: extremely local). Widespread, very local.

♂ N Greece

Description • Ups gc bluish black; small, submarginal bluish spots characteristic.

Flight-period • Northern range: mid-June to early August in one brood. Mediterranean region: mid-May to June & mid-July to August in two broods.

Habitat • Dry, rocky, or grassy places in scrubland, open woodland, or at woodland margins. LHPs: common honeysuckle (*Lonicera periclymenum*), *L. etrusca*, *L. implexa*, fly honeysuckle (*L. xylosteum*), alpine honeysuckle (*L. alpigena*), *L. nummulariifolia*, & perfoliate honeysuckle (*L. caprifolium*).

Conservation • In afforested regions, risks to habitat are same as those for *L. camilla*.

♂ N Greece

Limenitis camilla **White Admiral**

Distribution • N Spain, C Italy, & C N Greece (very local & rare), to S England, E Denmark, & Estonia. 0–1500 m. Widespread & locally common in most regions except those near limits of range. Apparently extinct in S Sweden.

Description • Ups gc dark chocolate brown, lacking blue submarginal spots (cf. *L. reducta*).

Flight-period • Mid-June to mid-August in one brood.

Habitat • Sunny clearings in large, often damp & humid, mature deciduous woodlands or pine forests with deciduous, wooded margins. LHPs: honeysuckle (*Lonicera periclymenum*), perfoliate honeysuckle (*L. caprifolium*), fly honeysuckle (*L. xylosteum*), & *Symphoricarpus racemosa*.

♂ N France

Behaviour • Graceful & gliding flight in dappled sunlight, usually settling on sunlit ferns, leaves of bushes, or bramble, the blossom of which greatly attracts both sexes.

Conservation • Potentially very vulnerable to damage or destruction of bushy/scrubby areas of forests by logging operations – often eliminated by tree-felling or log-stacking. Has extended range dramatically in S England since about 1930.

Hypolimnas misippus **False Plain Tiger**

Distribution • First reported from Canary Islands (Tenerife) in 1895, this powerful African migrant appears to have been resident on Gomera since 1987. Vagrant specimens recorded in Azores.

Description • Large; fw length 30 mm. Male ups very dark brown, with large, white, round discal spots, with additional upf subapical white oval spot – all spots with brilliant violet reflections at distal borders; uns gc brownish with ups pattern of white spots repeated but lacking violet reflections. Female is close mimic of both white & orangey-brown forms of female *Danaus chrysippus*.

Flight-period • Data limited. Observations relate to October–February.

Habitat • Records relate to gardens in coastal regions. LHP(s) on Gomera unknown; elsewhere, a wide variety of plant families, including species of *Ipomoea*, *Amaranthus*, *Hibiscus*, & *Sedum*, known to occur on Gomera.

Neptis sappho **Common Glider**

Distribution • NE Italy to E & S Balkans & C N Greece. 200–1550 m. Sporadic & generally very local, especially in SE range.

Description • Upf with prominent white stripe through cell; uph with white submarginal band broken by veins (cf. *N. rivularis*).

Flight-period • Mid-May to late June & July–August in two broods.

♂ N Greece

Habitat • Damp deciduous woodland, usually in river valleys. LHPs: spring pea (*Lathyrus vernus*) & black pea (*L. niger*).

Behaviour • Characteristic gliding flight is confined mostly to dappled shade of woodland canopy, where both sexes often rest or bask with open wings.

♂ N Greece

Neptis rivularis **Hungarian Glider**

Distribution • N Italy & SE Switzerland to S Poland, C & S Balkans & C N Greece. 500–1600 m. Very sporadic & local, but often common.

Description • Upf with indistinct, pale, broken line in cell; uph lacking submarginal white band (cf. *N. sappho*).

♂ N Greece

♂ N Greece

Flight-period • Late May to early August in one brood.

Habitat • Open deciduous or coniferous woodland. LHPs: elm-leaved spiraea (*Spiraea chamaedryfolia*), willow spiraea (*S. salicifolia*), *S. hypericifolia*, goatsbeard (*Aruncus dioicus*), & meadowsweet (*Filipendula ulmaria*).

Nymphalis antiopa **Camberwell Beauty**

Distribution • N Iberia, through most of Europe to N Fennoscandia, S Greece, & European Turkey. Distinction between resident colonies & migrants very difficult to establish with certainty in northern range, perhaps seasonally variable. Rare migrant in Britain. Absent from Atlantic islands. Recorded from Mediterranean islands of Corfu & Spetses. Generally 0–2000 m, occasional as migrants above 2600 m. Widespread, but usually encountered as single specimens.

Description • Ups gc dark maroon; yellow or creamy marginal borders preceded by blue spots.

Flight-period • One brood. S Europe: mid-June to July. Scandinavia: August–September. Hibernated specimens reappear March–June according to locality.

♂ Switzerland

Habitat • Open woodland, especially river valleys with an abundance of LHPs: willows, sallows (*Salix*), & poplars (*Populus*).

Behaviour • Powerful & well-known migrant, dispersing from breeding grounds soon after emergence. Strongly attracted to fermenting fruit & blossom of willow trees in spring. Hibernates as adult in cool, dark places, e.g., hollow trees, wood-stacks, road drains.

Nymphalis polychloros **Large Tortoiseshell**

Distribution • Iberia (widespread & common), through much of Europe, including most larger Mediterranean islands, to SE Sweden, Estonia, Greece & European Turkey. Occasional in England, Denmark, S Norway, & S Finland as a migrant. Possibly extinct in Britain as a resident. 0–1700 m.

Description • Ups gc yellowish brown; inner margin well-defined; legs & palpi dark brown or black (cf. *N. xanthomelas*).

♂ NE Greece

Flight-period • Late June to August in one brood. Hibernated specimens reappear March–April.

Habitat • Hot/dry, cool/damp woodland or bushy places, suitability of which seems to

♂ NE Greece

be determined largely, if not entirely, by presence of LHPs, especially elm (*Ulmus*), sallows & willows (*Salix*), & poplar (*Populus*).

Behaviour • Powerful in flight, as befits its marked dispersive–migratory tendency. Following emergence, often remains in large numbers & close company within breeding ground, feeding almost constantly on nectar-rich plants – giving striking impression of 'fuelling-up' for long journey. (Other large, migrant butterflies, e.g., Large White & Black-veined White, behave similarly.) Fond of basking in late afternoon sun. Hibernates as adult in cool, dark places, e.g., hollow trees, wood-stacks, road drains.

Conservation • Given continued availability of seemingly suitable habitats, progressive decline in NW range in recent decades more likely due to climatic change. Most hibernating butterflies are well-adapted to very low winter temperatures, but far less tolerant of damp conditions, most especially at higher temperatures.

♀ N Greece

Nymphalis xanthomelas
Yellow-legged Tortoiseshell

Distribution • Resident in C E Balkans & NW Greece (Varnous Mts). European records beyond this range more probably relate to migration, & possibly establishment of temporary breeding colonies. 0–2000 m. Generally widespread; locally often common but increasingly sporadic near limits of European range.

Description • Resembles *N. polychloros*. Ups gc bright, more reddish; inner edge of dark submarginal borders suffused – indistinct; legs & palpi light brown or buff.

Flight-period • July–September in one brood. Hibernated specimens reappear May.

Habitat • Deciduous woodland, especially in river valleys containing abundance of LHPs: willows & sallows (*Salix*), & poplar (*Populus*).

Behaviour • Shows strong migratory tendency. Hibernated specimens strongly attracted to nectar of willow blossom.

Nymphalis vaualbum **False Comma**

Distribution • Resident in C E Balkans. Records beyond this range more probably relate to migration, although distinction between permanent populations, migration, & temporary colonies established through migration very difficult to establish with reliability. 0–1400 m. Widespread but becoming increasingly uncommon.

Description • Resembles *N. polychloros* & *N. xanthomelas* but with white markings near upf apex & uph costa – prominent & distinctive.

Flight-period • June–July in one brood. Hibernated specimens reappear March–April.

Habitat • Clearings in deciduous woods, especially river margins with an abundance of LHPs: willows & sallows (*Salix*), poplar (*Populus*), & elms (*Ulmus*).

Behaviour • Shows strong migratory tendency.

Inachis io **Peacock Butterfly**

Distribution • Most of Europe, including Mediterranean islands of Sicily, Corsica, Sardinia, Corfu, Samos, & Simi. 0–2500 m. Widespread & common.

Description • Upf & uph with large, distinctive eye-spots – quite unlike any other European butterfly.

Flight-period • Normally June–August but enters hibernation late July to early October, according to weather conditions. Hibernated specimens reappear March–May.

♂ N France

Habitat • Open, sunny places in a wide variety of habitat types, including sheltered, rocky gullies at higher altitudes. LHPs: principally common stinging nettle (*Urtica dioica*); also, wall-pellitory (*Parietaria officinalis*).

Behaviour • Shows some dispersive tendency. Hibernates as adult in cool, dark places, including wood-stacks, outhouses, & stone walls. Hibernated adults often feed on blossom of goat-willow (*Salix caprea*) in spring. Frequent visitor to 'butterfly-bush' (*Buddleia davidii*) in summer.

Vanessa atalanta **Red Admiral**

Distribution • Azores. Canary Islands. Most of Europe, including all larger & many smaller Mediterranean islands, as a resident or migrant. In northernmost range (Britain, & most of Fennoscandia) appearance probably due largely to migration. 0–2500 m.

Description • Ups gc brownish black; upf subapical spots white, with bright, unbroken, orangey-red band from costa, through cell to anal angle (cf. *V. indica*); uph orangey-red band with four black spots in s2–5.

♂ S Greece

♂ S Greece

Flight-period • June–October in one brood. Hibernated specimens usually reappear March–April, but recorded on warm days in January–February near Mediterranean coast.

Habitat • Occurs in many habitat types, but usually those with an abundance of LHP in open, sunny situations. LHP: principally common stinging nettle (*Urtica dioica*); also annual nettle (*U. urens*) & pellitory (*Parietaria*).

Behaviour • Common migrant. Hibernates as adult (sometimes successfully in Britain) in cool, dark places, including wood-stacks & outhouses. Strongly attracted to nectar-rich plants, especially bramble (*Rubus fruticosus*), hemp agrimony (*Eupatorium cannabinum*), ivy (*Hedera helix*), ice-plant (*Sedum spectabilis*), & 'butterfly-bush' (*Buddleia davidii*).

Vanessa indica **Indian Red Admiral**

Distribution • Madeira. Porto Santo as an occasional migrant. Resident in Canary Islands except Lanzarote & as an infrequent migrant on Fuerteventura. 0–1500 m. Locally very common.

Description • Resembles *V. atalanta*, but upf white markings greatly reduced, orangey-red band wider, irregular, containing conspicuous dark spot near wing-base.

Flight-period • Continuously brooded: recorded in all months.

Habitat & behaviour • Laurel forests, but often found in urban areas in search of nectar-rich plants. LHPs: nettles (*Urtica*) & pellitory (*Parietaria*).

Vanessa cardui **Painted Lady**

Distribution • Throughout nearly all of Europe to well within Arctic Circle as migrant. Resident of Canary Islands, Madeira, & probably warmest regions of Mediterranean. Relatively rare in Ireland & Scotland. Reported rarely from Iceland, which has no indigenous butterflies. 0–3000 m; commonest below 1500 m, with migrants accounting for most observations at highest altitudes. Often occurs in great abundance in Mediterranean districts in early spring, following establishment of coastal colonies from N African migrants.

Description • Resembles *V. virginiensis* but upf lacking white pd spot in s2 & blue pupils in pd black spots on uph.

♂ N Belgium

Flight-period • Continuously brooded in regions of permanent residence. Northern Europe: migrants usually appear May–June; breeding & further dispersal persists until onset of cold weather.

Habitat • Perhaps best known & most wide-ranging of all European migrant butterflies. Many habitat types containing an abundance of LHPs; these include a wide range of plant families, especially thistles (*Cirsium* & *Carduus*).

Behaviour • Strongly attracted to thistles, knapweeds (*Centaurae*) & other nectar-rich plants & shrubs. Common summer visitor to 'butterfly-bush' (*Buddleia davidii*) in gardens & parks.

Cynthia virginiensis **American Painted Lady**

Distribution • Canary Islands (reportedly resident on Tenerife; apparently extinct on Gomera & La Palma). Occasional migrant in Azores, W & S Portugal, more rarely in Spain. Not reported from Madeira. 0–1500 m. Extremely local.

Description • Resembles *C. cardui* but ups gc lacking pinkish tone; upf with additional white pd in s2; uph pd spots in s2 & s5 with blue pupils; unh spots in s2 & s5 in well-defined, brownish pd band, large & conspicuous.

Flight-period • Recorded in all months on Tenerife except February & November. Most probably hibernates in winter months.

Habitat • Flowery places. LHP on Tenerife unknown.

Aglais urticae **Small Tortoiseshell**

Distribution • Most of Europe. Absent from Atlantic islands, & Mediterranean islands except Sicily, Corsica, & Sardinia. 0–3000 m. Widespread & common.

Description & variation • Ups gc red with prominent yellow & black markings on fw; uph basal area dark brown. § At high altitudes. Generally larger; ups gc brighter. § Corsica & Sardinia (700–2500 m), f. *ichnusa*. Upf black spots in s2 & s3 greatly reduced or absent.

Flight-period • May–October in 1–3 broods, according to altitude & locality. Hibernated specimens reappear March–April.

♀ C Greece

Habitat • Almost all sites containing LHP: common stinging nettle (*Urtica dioica*). Especially common in rural areas of human habitation where soil disturbance & enrichment encourage establishment of 'nettle-beds'.

Behaviour • Fond of basking on hot walls, paths, etc. Commonly observed at mountain summits. Frequent visitor to 'butterfly-bush' (*Buddleia davidii*), ice-plant (*Sedum spectabilis*), & Michaelmas daisy (*Aster novi-belgii*) in summer & early autumn. Hibernation sometimes disrupted on warm winter days.

Polygonia c-album Comma Butterfly

Distribution • Most of Europe, including England, Wales, & Mediterranean islands of Sicily, Corsica, Sardinia, & Corfu. 0–2000 m. Widespread & common.

Description & variation • Unh cell-end with white 'comma' (cf. *P. egea*). Early summer brood (f. *hutchinsoni*): uns gc yellowish brown with bright, variegated pattern. Late summer–autumn brood; uns dark brown with obscure, dark greenish marbling.

♀ N Greece

Flight-period • Late May to October in 1–3 broods, according to locality & altitude. Hibernated specimens reappear March–April.

Habitat • Mainly deciduous woodland clearings containing LHPs: especially common stinging nettle (*Urtica dioica*) & goat willow (*Salix caprea*); also white willow (*S. alba*), hop (*Humulus lupulus*), elm (*Ulmus*), & currant family (*Ribes*).

Behaviour • Hibernation sometimes disrupted briefly on warm days in late winter. Adults often feed on willow blossom in early spring.

♂ E Spain (summer brood)

Polygonia egea **Southern Comma**

Distribution • Mediterranean region from SE France, through Corsica & Sicily to S Balkans & Greece, including Corfu, Crete, & many other Aegean islands. 0–1700 m, generally below 1100 m. Sporadic & local.

Description • First brood: ups yellowish orange with variable dark brown pd & discal spots; uns gc yellowish with complex darker, variegated pattern; unh with small, white y-shaped mark at cell-end. Later brood(s): ups & uns gc usually slightly darker, markings heavier.

♂ S Greece

Flight-period • May–October in 2–3 broods according to locality. Hibernated specimens reappear late March.

Habitat • Hot, dry, steep, rocky places; also old stone walls providing a convenient habitat for LHP: wall-pellitory (*Parietaria officinalis*).

Behaviour • Especially fond of basking on rock faces, walls, hot paths, etc.

Araschnia levana **Map Butterfly**

Distribution • N Spain (extremely local in E Pyrenees), through N & C France, Switzerland to E Denmark, S Sweden (SW Skåne) & Estonia to Bulgaria & NE Greece (extremely sporadic & local). 0–1400 m. Generally sporadic & local, especially near limit of range.

Description & variation • First brood: ups gc dusky orange with complex pattern of black markings; uns with complex variegation well-delineated by veins – vaguely

resembling a map & giving rise to common name. Second brood: ups gc dark brown, with creamy-yellow or white discal band disrupted at v4 upf.

Flight-period • May–June & July–August in two broods.

♂ E France (1st brood)

Habitat • Deciduous woodland clearings or bushy margins, usually containing an abundance of LHP: common stinging nettle (*Urtica dioica*).

Behaviour • Often basks in sheltered, sunny positions on ferns or leaves of shrubs or lower branches of trees.

♂ C France (spring brood)   ♂ NE Greece (summer brood)

Argynnis pandora **Cardinal Fritillary**

Distribution • Canary Islands (La Palma, Gomera, Hierro, & Tenerife). 500–1500 m. S Europe, including most Mediterranean islands, to C E Balkans. 50–1650 m. Widespread but local.

Description • Ups dusky yellowy orange gc heavily suffused fulvous-grey on fw & olive green on hw; black markings very bold: unf anterior area pink.

Flight-period • Generally mid-May to early July in one brood; possibly double-brooded in some localities of S Europe. Canary Islands: late May to mid-September in a prolonged emergence of 2+ broods.

Habitat • Open, bushy clearings in deciduous or pine woodland. LHPs: violets (*Viola*).

Behaviour • Strongly attracted to nectar-rich plants, especially thistles (*Carduus* & *Cirsium*) & knapweeds (*Centaureae*).

♀ NE Greece

♀ NE Greece

♂ NE Greece

Argynnis paphia **Silver-washed Fritillary**

Distribution • Most of Europe, including Ireland, S Britain, Sicily, Corsica, Sardinia, Elba, Lesbos, Samos, Ikaria, & Andros. 0–1500 m. Widespread, locally common.

♀ NE Greece

♀ NE Greece

Description & variation • Male upf with distinctive sex-brand along veins. Both sexes: unh gc & development of 'silver-wash' variable in Mediterranean region. § Female f. *valezina*. Ups suffused olive green, with more usual black markings replaced by dark brown (a recurrent form in most European populations, including England).

Flight-period • Late May to September in one brood.

Habitat • Woodland clearings or borders with bushy margins. LHPs: violets (*Viola*).

Behaviour • Strongly attracted to nectar-rich plants, especially bramble (*Rubus fruticosus*), thistles (*Carduus* & *Cirsium*), & knapweeds (*Centaureae*).

Argynnis laodice **Pallas' Fritillary**

Distribution • Latvia. Lithuania. E Poland. Slovakia.
E Hungary. Romania. An occasional migrant in Estonia, SE
Sweden, & S Finland. Occurrence in northern & western range
uncertain. Altitudinal range uncertain; generally below 250 m.

Description • Male ups bright orange or fulvous with dark
markings & variable black suffusion; unh basal & discal areas
yellowish green with thin, brown mediobasal & discal lines,
separated from pd area by white striae; buff pd area with
variable pinkish-lilac suffusion. Female similar; ups paler, with
small but conspicuous white subapical marking.

Flight-period • July–August in one brood.

Habitat • Damp, flowery meadows in open woodland. LHP:
marsh violet (*Viola palustris*).

Behaviour • Shows marked dispersive–migratory tendency.
Both sexes attracted to bramble blossom.

Argynnis aglaja **Dark Green Fritillary**

Distribution • Widespread & common in most of Europe
including Britain, Ireland, Orkney Islands, & European
Turkey. Absent from Mediterranean islands except Sicily &
Evia (Mt Dirfis). 0–2200 m.

♂ N Greece

Description & variation • Unh greenish
suffusion & silvery-white markings
characteristic – lacking unh reddish-brown
pd spots (cf. *A. adippe* & *A. niobe*). In
colder climates, e.g. Scotland, ups black
markings are often heavy, sometimes with
extensive black suffusion, at times almost
obliterating gc.

Flight-period • June–August in one
brood.

Habitat • Open, grassy, flowery slopes;
clearings in light woodland; damp meadows;
heaths; moorlands. LHPs: violets & pansies,
including hairy violet (*Viola hirta*), wild
pansy (*V. tricolor*), & marsh violet
(*V. palustris*).

♀ N Greece

Argynnis adippe High Brown Fritillary

Distribution • Most of Europe, including England & Wales. Absent from Mediterranean islands except Sicily & Evia (Mt Dirfis). 0–2100 m. Generally widespread but more sporadic & local in northern range.

♀ NW Greece

♂ NW Greece

Description & variation • Male upf sex-brands on v2 & v3 conspicuous; uph with hair-fringe along v7 (cf. *A. niobe*); unh reddish brown, silver-pupilled pd spots characteristic (cf. *A. aglaja*). § E Pyrenees to S Balkans & Greece. Unh gc buff or yellowish, without normal, green suffusion; all silver markings lacking except small pupils of pd reddish-brown spots. § Iberia. Unh heavily suffused olive green in discal & basal areas; all silver spots prominent.

Flight-period • Late May to August in one brood.

Habitat • Dry, grassy, bushy places, usually in woodland clearings. LHPs: violets (*Viola*).

Conservation • Increasingly scarce in NW regions, including Britain. Whilst changes in woodland management are widely held responsible, influence of climatic change is difficult to isolate. Decline or disappearance, or, indeed, increase in ranges of other species, difficult to explain entirely, if at all, on basis of habitat change or loss.

♂ E Spain

Argynnis niobe **Niobe Fritillary**

Distribution • Most of Europe from Mediterranean coast to S Fennoscandia. Absent from Mediterranean islands except Sicily. 0–2400 m. Widespread, locally common.

Description & variation • Male ups resembles *A. a. adippe*. Upf sex-brand on v2 & v3 narrow, sometimes absent; unh usually with small yellow or silver spot, often with black pupil or thin black ring, near cell-base – a useful distinguishing character (cf. *A. adippe* & *A. aglaja*). § F. *eris*. Unh silver spots absent, but usual positions outlined black; gc variable, pale greenish yellow to yellowish buff. Replaces nominate form, partly or completely, in many regions.

♀ Samos, Greece

Flight-period • Late May to August in one brood.

Habitat • Open, grassy places or rocky gullies, often in scrub or woodland clearings. LHPs: violets (*Viola*).

♀ Samos, Greece

Argynnis elisa Corsican Fritillary

Distribution • Corsica. Sardinia. 800–1500 m. Widespread, locally common.

Description • Resembles a small version of *A. adippe*, a species which does not occur on Corsica or Sardinia.

Flight-period • Late June to mid-August in one brood.

Habitat • Dry, open heaths, bushy places, or light woodland. LHP: Corsican violet (*Viola corsica*).

Issoria lathonia Queen of Spain Fritillary

Distribution • Madeira & Canary Islands. Most of Europe to 63°N in Fennoscandia as resident, & 66°N as migrant in W Finland. Rare as migrant in S England. Resident on Corsica, Sardinia, & Sicily. Not reported from E Mediterranean islands. Generally, 0–2700 m but vagrants/migrants occur at higher altitudes. Commoner in summer months as populations expand & disperse.

Description • Ups bold, black markings & unh large, silver spots characteristic; upf margin concave below subapex.

Flight-period • March–October in three broods.

Habitat • Almost all habitat types. LHPs: pansies & violets (*Viola*).

Behaviour • Well-known migrant. In late afternoon sun, adults often bask on hot paths, walls, etc.

♂ C Greece

♂ C Greece

Brenthis hecate **Twin-spot Fritillary**

Distribution • Spain, through S France & Italy to C Balkans, European Turkey, & C Greece. 25–1500 m, generally above 500 m. Very sporadic & local, especially in western range.

Description • Ups & unh black, submarginal & pd spots in uniform series; unh black marginal lines well-defined (cf. *B. daphne* & *B. ino*).

Flight-period • Generally late May to late July in one brood. Sometimes emerges in April in low-lying Mediterranean sites.

Habitat • Flowery, grassy meadows, usually sheltered by bushes or light woodland. LHPs: meadowsweet (*Filipendula ulmaria*) & dropwort (*F. vulgaris*).

♂ NE Greece

Brenthis daphne **Marbled Fritillary**

Distribution • Spain, through S France & S Poland to Balkans & C Greece. Absent from Mediterranean islands except Sicily & Samos. 75–1750 m. Widespread, locally common.

Description • Ups pd spots irregular in size (cf. *B. hecate* & *B. ino*); unh, base of s4 (adjacent to cell-end) yellow, partly shaded, or striated orangey brown (cf. *B. ino*).

Flight-period • Late May to early August in one brood.

Habitat • Bushy, flowery places, often in woodland clearings containing an abundance of LHPs: bramble (*Rubus fruticosus*) & raspberry (*R. idaeus*).

♂ N Greece

♂ N Greece

Brenthis ino **Lesser Marbled Fritillary**

Distribution • N & E Spain to Fennoscandia & Balkans. In Italy, restricted to foothills of S Alps & Calabria. Absent from Mediterranean islands. 0–2000 m. Sporadic & local.

Description & variation • Unh base of s4 (adjacent to cell-end) wholly yellow (cf. *B. daphne*). Ups black markings & dark suffusion very variable, especially in Scandinavia.

Flight-period • June–August in one brood. Emergence date depends on locality.

Habitat • Damp, flowery, sometimes marshy places, sheltered by scrub or woodland. LHPs include meadowsweet (*Filipendula ulmaria*), cloudberry (*Rubus chamaemorus*), raspberry (*R. idaeus*), & goatsbeard (*Aruncus dioicus*).

Boloria pales **Shepherd's Fritillary**

Distribution • On most higher mountains from N & E Spain, through C Europe to Balkans. Not reported from Greece. Generally 1500–2800 m, occasionally 1200 m. Local, but usually very common.

♂ S Switzerland

♂ E Switzerland

Description & variation • Upf black discal markings bold, macular (cf. *B. napaea*); unf markings obscure – 'ghosted'; unh basal & pd areas reddish. § SE France, S Switzerland, & C Apennines to Balkans, f. *palustris*. Smaller; ups paler; upf discal markings thinner; unh red markings more extensive. § Cantabrian Mts & Pyrenees, f. *pyrenesmiscens*. Larger; ups gc slightly paler; uph black basal suffusion reduced, not encroaching cell. Female ups sometimes suffused greyish violet.

Flight-period • Late June to early September in one brood. Emergence date depends on locality, especially altitude.

Habitat • Flowery alpine/subalpine meadows. LHP: long-spurred pansy (*Viola calcarata*).

Boloria napaea **Mountain Fritillary**

Distribution • Very local in E Pyrenees. C Alps from France to Austria. 1500–2500 m. Norway & W Sweden from 60°N to North Cape. 0–1100 m.

Description • Male ups paler; upf black discal markings thin, not macular (cf. *B. pales*); unh paler. Female ups suffused grey, often with strong violet reflections & greenish basal overtones, most noticeable in fresh specimens.

Flight-period • Late June to August in one brood.

Habitat • S Europe: flowery alpine/subalpine meadows. Scandinavia: similar, but at much lower altitudes, occasionally sea-level at highest latitudes. LHPs: yellow wood violet (*Viola biflora*) & alpine bistort (*Polygonum viviparum*).

Boloria aquilonaris **Cranberry Fritillary**

Distribution • S France (Massif Central) to Austria & Fennoscandia. Very sporadic & local in W Europe, becoming progressively less so from S Germany to Baltic countries. Widespread & locally common in Fennoscandia. 100–2000 m.

♂ N Norway

♂ N Norway

Description • Male ups gc bright, often fiery red. Female similar but gc less intense & often with dusky suffusion.

Flight-period • Mid-June to August in one brood.

Habitat • Peat bogs or wet heaths often sheltered by light woodland, usually near permanent water. Distinctive habitat, not occupied by similar species. LHP: cranberry (*Vaccinium oxycoccos*).

Conservation • As with all other wetland species, habitats especially vulnerable to drainage in adjacent areas.

Boloria graeca **Balkan Fritillary**

Distribution • SW Alps of France & Italy. C Balkans to C Greece. Local but generally very common. 1450–2600 m.

Description & variation • Hw sharply angled at v8; unh pd ocellated spots well-defined. § SW Alps, f. *tendensis*. Unh ocellated pd spots generally better developed.

♂ C Greece

♂ C Greece

Flight-period • Mid-June to early August in one brood.

Habitat • Open grassland, often near mountain summits. Many colonies contain prostrate juniper (*Juniperus communis nana*). LHP: pansies (*Viola*).

♀ NW Greece

♀ C Greece

Proclossiana eunomia Bog Fritillary

Distribution · Very sporadic & local in W, C, & SE Europe. 300–1900 m. Widespread & common in Norway, Sweden, & Finland; sporadic but locally common in Baltic states. 50–900 m.

♂ N Norway

♂ N Norway

Description & variation · Ups gc yellowy orange; black markings uniform; unh pd spots well-defined – quite characteristic. § NE Europe, f. *ossiana*. Generally smaller; ups & uns markings well-defined; unh basal, discal, & pd spots white or silver; ups often with dark suffusion, especially in female.

Flight-period · One brood. C Europe: late May to early July. NE Europe: mid-June to mid-July.

Habitat · Boggy or marshy places, often by rivers or lakes. S Europe LHP: bistort (*Polygonum bistorta*). NE Europe: alpine bistort (*Polygonum viviparum*) & marsh violet (*Viola palustris*).

Conservation · Most habitats, which in southern part of European range are very few & far between, extremely small & correspondingly vulnerable. Many have been destroyed by drainage & afforestation.

♀ S Belgium

Clossiana euphrosyne
Pearl-bordered Fritillary

Distribution • From N Spain, through most of Europe, including Britain & C W Ireland, to North Cape & S Greece. Absent from Mediterranean islands except N Sicily. 0–1900 m. Widespread, locally common.

♂ NW Greece

Description & variation • Superficially resembles *C. selene* but unh brick-red markings & prominent silver spots distinctive. § Fennoscandia. Generally smaller; ups markings with variable dark suffusion, sometimes obscuring gc, especially in female.

Flight-period • N Europe & at high altitudes in S Europe: late May to July in one brood. Elsewhere: April–June & July–September (partial second brood may occur).

Habitat • Woodland clearings. LHPs: violets (*Viola*).

♂ NW Greece

♀ N Norway

Clossiana titania **Titania's Fritillary**

Distribution • France (Massif Central). C Alps to Finland, SW Serbia, Romania, & S Bulgaria. Distribution greatly disrupted in eastern regions. 300–1800 m. Locally common.

Description & variation • Male ups brownish orange; black markings fine; unh marbled yellow & brown with greenish or violet tints. Female ups & uns paler. § C & S Switzerland to Finland & Balkans, f. *cypris*. Male ups reddish; all black markings heavy; unh darker, marbled brown, purple, maroon, pale violet, & yellow, giving an overall very striking & distinctive appearance. Female ups gc paler. Resembles *C. dia* but larger.

Flight-period • Mid-June to late August in one brood.

Habitat • Flowery meadows in woodland clearings. LHP: bistort (*Polygonum bistorta*).

Clossiana selene
Small Pearl-bordered Fritillary

Distribution • Throughout most of Europe except Ireland & warmer districts of Mediterranean countries. 0–2200 m. Locally common.

Description & variation • Resembles *C. euphrosyne* but unh silver spots less prominent & reddish markings darker; basal black spot usually larger. § Fennoscandia, f. *hela*. Generally smaller with variable dark ups suffusion.

Flight-period • Higher latitudes & altitudes: mid-May to early July in one brood. Warmer regions: early May to late June & mid-July to early September (partial second brood has occurred in England in exceptionally warm summers).

Habitat • Forest clearings, often in damp, flowery, & grassy places; sheltered marshy sites in Arctic regions. LHPs: violets (*Viola*).

♂ S Belgium

♀ S Belgium

Clossiana chariclea **Arctic Fritillary**

Distribution • Lappland. 100–1400 m. Very local but often abundant.

Description • Unh prominent silver markings, reddish-brown pd area enclosing dark spots characteristic.

♂ N Norway

Flight-period • June to early August in one brood. Emergence date (usually late June to early July) much dependent on weather conditions.

Habitat • Sheltered sites, usually with low shrubs, on open rocky tundra. LHP not reported; ova sometimes laid on dead vegetation.

Behaviour • Often rests with open wings in shelter of stones or shrubs, even in warm, overcast conditions.

♀ N Norway

♀ N Norway

Clossiana freija **Frejya's Fritillary**

Distribution • Norway, Sweden, & Finland. Widespread, generally common. Possibly now extinct in Baltic countries. 200–1000 m.

Description • Ups irregular black discal markings bold; unh 'jagged' discal line distinctive.

Flight-period • Generally late May to late June in one brood. May appear late July to early August in retarded seasons.

♂ N Norway

Habitat • Wet or dry grassy or rocky places, usually with some shelter from shrubs or trees. LHPs include whortleberry (*Vaccinium uliginosum*), cloudberry (*Rubus chamaemorus*), & bearberry (*Arctostaphylos uva-ursi*).

Clossiana dia **Weaver's Fritillary**

Distribution • N Spain, through NW France to Baltic countries, Balkans, European Turkey, & C Greece. Absent from Mediterranean islands except N Sicily. 500–1550 m. Widespread but local.

Description • Resembles *C. titania* but smaller; hw more sharply angled at v8.

Flight-period • Late April to early September in 2–3 broods.

♀ N Greece

Habitat • Dry or damp, bushy, grassy, & flowery woodland clearings. LHPs: violets & pansies, including sweet violet (*Viola odorata*) & wild pansy (*V. tricolor*).

Behaviour • A frequent visitor to bramble blossom.

♀ N Greece   ♂ E Spain

Clossiana polaris **Polar Fritillary**

Distribution • Lappland. 100–1400 m. Extremely local & generally very uncommon.

Description • Ups dull yellowy orange, often with smoky suffusion in female; unh markings very distinctive, especially shape of white marginal spots.

Flight-period • Late May to early August in one brood. Emergence date (usually early July) much dependent on weather conditions.

Habitat • Sheltered hollows or slopes on open, rocky tundra. LHP(s) unconfirmed in Europe; possibly mountain

avens (*Dryas octopetala*) &/or bog wortleberry (*Vaccinium uliginosum*).

Behaviour • Often rests with open wings amongst shelter of stones or low shrubs, even in dull, warm conditions. Flies rapidly & close to ground.

♂ N Norway

♂ N Norway   ♀ N Norway

Clossiana thore **Thor's Fritillary**

Distribution • C Alps. 800–1800 m. Norway, W Sweden (62–70°N) & NW Finland. 300–1000 m. Sporadic & local.

Description & variation • Ups orange with bold, dark brown markings; gc in basal areas often largely obscured by dark suffusion. § N Scandinavia, f. *borealis.* Ups gc slightly paler; ups black markings much reduced; unh paler.

♂ N Norway

Flight-period • Mid-June to early August in one brood. In Lappland, emergence date depends on weather conditions.

Habitat • Shaded woodland clearings, often bordering rivers or small streams.

Behaviour • Both sexes often bask on flower heads in dappled sunlight. LHPs: violets, including yellow wood pansy (*Viola biflora*).

♂ N Norway

♀ E Switzerland

Clossiana frigga **Frigga's Fritillary**

Distribution • Most of Norway & Sweden from 60°N to North Cape. Generally local & uncommon, more widespread north of Arctic Circle. Widespread in Finland. Very rare & local in Baltic states, possibly now extinct in Latvia. 100–450 m.

♂ N Norway

Description & variation • Ups gc sometimes very pale; black markings generally bold but variable; unh basal & discal gc pale reddish brown to maroon with pinkish or violet overtones.

Flight-period • Late June to late July in one brood.

Habitat • Wet places, usually sheltered by trees & scrub, often near ponds. LHP: cloudberry (*Rubus chamaemorus*).

♂ N Norway

♀ N Sweden

Clossiana improba **Dusky-winged Fritillary**

Distribution • Fennoscandia north of 68 °N. 600–1050 m. Sporadic & very local.

Description • Ups gc dusky orangey brown; usual markings for genus are complete but indistinct due to extensive greyish-brown suffusion; unh pale dusky brown; white marks in s4 & s7, & narrow white costal line clearly defined.

Flight-period • Late June to early August in one brood.

Habitat • Open, grassy terrain, usually in lee of slopes or ridges affording shelter from prevailing NW winds. LHP(s) unknown in Europe; possibly willows (*Salix*), of which two European species, arctic willow (*S. arctica*) & netted willow (*S. reticulata*), are known host plants in N America.

Behaviour • Flies fast & very close to ground when disturbed. Well-camouflaged when settled with open wings on similarly coloured patches of bare ground. Both sexes especially fond of nectar of moss campion (*Silene acaulis*). Males sometimes take moisture from damp, peaty soil.

Melitaea cinxia **Glanville Fritillary**

Distribution • N Portugal & Spain, through most of Europe, including S England (Isle of Wight only), Channel Islands, Sicily, & Corfu, to S Fennoscandia, Balkans, & Greece. 0–2000 m. Generally widespread & very common.

Description • Ups black markings regular; uph pd black spots distinctive. Female ups gc paler yellowish buff, sometimes with smoky or greyish-green suffusion; unh pattern of markings distinctive, resembling *M. arduinna* but discal band paler, usually white.

♂ NW Greece

Flight-period • Most regions: late April to early August in one prolonged emergence. Some SW regions: May–June & August–September in two broods.

Habitat • Warm, grassy, flowery places, including meadows, open hillsides, woodland clearings, & margins of cultivation. LHPs: plantains, especially ribwort plantain (*Plantago lanceolata*); also knapweeds (*Centaurea*) & speedwells (*Veronica*).

Melitaea arduinna **Freyer's Fritillary**

Distribution • Romania. Republic of Macedonia. Bulgaria. NW Greece. 500–1500 m. Very sporadic & local, sometimes common.

Description & variation • Superficially resembles *M. cinxia* but quite distinct in overall character. Female ups dark markings variable.

♂ NW Greece

Flight-period • Late May to early August in one brood.

Habitat • Grassy, flowery places sheltered by bushes or woodland. LHP: Greek knapweed (*Centaurea graeca*).

♂ NW Greece

♀ NW Greece

Melitaea phoebe Knapweed Fritillary

Distribution • Iberia to SE Latvia, Balkans, & Greece. Absent from Mediterranean islands except Sicily, Chios, Lesbos, Poros, Spetses, & Evia. 0–1900 m. Generally widespread & common; sporadic & very local in northern range.

Description & variation • Uph dark pd band well-defined (cf. *M. Aetherie*). Size, ups gc, & black markings, regionally, locally, & seasonally extremely variable. Later broods often smaller & paler in hot, low-lying localities. § Spain. Especially in second brood, ups orange submarginal band in striking colour contrast to paler marginal & discal markings.

Flight-period • Mid-April to early September in 2–3 broods, according to locality. Possibly single-brooded with prolonged emergence at high altitude.

Habitat • Dry, often very hot, open, flowery, & grassy places, usually amongst scrub or in light woodland. LHPs: a wide range of knapweeds (*Centaurea*).

♂ S Spain   ♀ S Spain

Melitaea aetherie Aetherie Fritillary

Distribution • S Portugal (Algarve; 25–250 m). S Spain (provinces of Huelva, Cadiz, Malaga & Jaen; 50–700 m). NW Sicily (Petralia, La Madonie, Ficuzza, & Lupo; 800–1100 m). Extremely sporadic & local but sometimes common.

Description • Male ups gc bright uniform reddish orange; pattern of black markings as for *M. phoebe* but greatly reduced, often absent in pd area; marginal & submarginal black markings complete & bold. Female ups dark markings heavier, sometimes with upf & posterior half of uph suffused greyish.

Flight-period • Iberia: mid-April to May in one brood. Sicily: May–June & September in two broods.

Habitat • Hot, dry, grassy, & flowery places, most often in neglected areas of cultivation. LHPs include knapweeds, especially *Centaurea calcitrapa* & *C. carratracensis*.

Conservation • Most colonies occur in areas of human activity & are thus very vulnerable. Coastal development/urbanization in S Portugal (Algarve) directly responsible for widespread, local extinction.

Melitaea didyma **Spotted Fritillary**

Distribution • Most of S & C Europe to S Belgium & SE Latvia. Reported from Sicily, Elba, Corfu, Levkas, Evia, Thassos, Limnos, Lesbos, Chios, & Samos. 0–2300 m. Widespread & generally common in Mediterranean region; sporadic & very local in northern range.

♂ S Greece

Description & variation • Resembles *M. trivia*. Unh black spots in marginal band usually rounded; orange discal band irregular but continuous. Ups gc & black markings, regionally, locally, & seasonally extremely variable, with altitude showing an additional & significant effect on gc & development of ups black markings, especially suffusion in female. Whilst many named forms have been described to account for local & geographical variation, overall picture is one of considerable complexity & no small confusion.

Flight-period • Mid-April to September in 2–3 broods.

Habitat • Dry, flowery places in a diverse range of habitat types, including areas of cultivation. LHPs include toadflax (*Linaria*), plantains (*Plantago*), speedwells (*Veronica*), & valerian (*Valeriana*).

♂ S Greece

♀ S Greece

Melitaea trivia **Lesser Spotted Fritillary**

Distribution • N Portugal & N & E Spain. 500–1200 m.
N Italy & Apennines. Slovenia to Slovakia, S Balkans, & Greece,
including Aegean islands of Thassos, Lesbos, Chios, Samos,
Ikaria, & Kos. 0–1700 m. Widespread & common in SE range,
very sporadic & local in Iberia & Italy.

♂ N Greece

Description & variation • Resembles
M. didyma. Unh black spots in marginal
band vaguely triangular, flattened near outer
margin; orange discal band irregular &
usually broken. Second brood: often small;
ups gc paler, black markings reduced. §
Iberia. Ups gc uniform pale yellowy orange;
black markings clearly defined.

Flight-period • Mid-April to early May &
June–August in two broods.

Habitat • Dry, often hot, flowery places,
often amongst scrub or in margins of
cultivated ground. LHPs: mulleins, especially
great mullein (*Verbascum thapsus*).

♂ S Greece

♀ N Greece

Melitaea diamina False Heath Fritillary

Distribution • N Spain, NW & S France, N Italy, & S Belgium, to S Fennoscandia & S Balkans. 100–2000 m. Sporadic & local, especially near limits of range.

♂ E France

♀ SW France

Description & variation • Male ups gc orange or fulvous on disc, pale yellow or white near outer margins; black markings extensive, obscuring much of gc, especially on hw; uns gc very pale; narrow, yellowish marginal band, bordered by fine black lines (cf. *M. britomartis*). Female ups gc paler. § Spain, Italy & elsewhere at lower altitudes. Ups dark suffusion absent, markings clearly defined; upf usually with well-defined, dumbbell-shaped discal mark in s1b.

Flight-period • Higher latitude & altitude: May–July in one brood. Foothills of S Alps & Spain: May–July & August–September in two broods.

Habitat • Damp, grassy, flowery places, in sparse, open deciduous or coniferous woodland or forest margins. LHPs: valerians, including common valerian (*Valeriana officinalis*) & marsh valerian (*V. dioica*).

♀ E Switzerland

Mellicta athalia **Heath Fritillary**

Distribution • Most of Europe, including SW & SE England (extremely local). Absent from Mediterranean islands, S Greece, & S Spain except Sierra Nevada. 0–2600 m. Generally widespread & common.

Description & variation • Unf pale marginal spots in s2 & s3 usually with conspicuous black internal borders. This character, & overall appearance, usually allows easy separation from closely related species. Size, gc, & markings regionally & locally variable.

Flight-period • Generally mid-May to mid-August in one prolonged brood. Partial second brood (mid-August to September) reported from some southern localities, possibly in error for extended emergence of one brood.

Habitat • Dry or damp, grassy, flowery places in a wide variety of habitat types. LHPs include ribwort plantain (*Plantago lanceolata*), germander speedwell (*Veronica chamaedryas*), common cow-wheat (*Melampyrum pratense*), rusty foxglove (*Digitalis ferruginea*), small yellow foxglove (*D. lutea*), & common toadflax (*Linaria vulgaris*).

♂ NW Greece

♂ S England

♂ S England

Mellicta deione **Provençal Fritillary**

Distribution • Iberia, through S France, W Switzerland, N Italy including Dolomites. 100–1600 m. Sporadic & local, especially in eastern range.

Description & variation • Male ups gc uniform yellowy orange; black markings fine; uns gc pale; unh yellowy orange bands bright, well-defined; submarginal band usually with distinctive rounded orangey-red spots in each space. Female ups gc paler, usually with distinctly paler shade in fw pd area.

♀ N Spain

These features, coupled with overall character, generally allow easy separation from similar species. Ups gc & black markings, with or without dark suffusion, regionally & locally variable.
   § W Switzerland (Rhône Valley) & Portugal. Populations with especially dark markings occur.

Flight-period • April to early September, generally in two broods. W Switzerland & NW Italy: May–July in two broods .

Habitat • Open, flowery, grassy places amongst bushes & scattered trees. LHPs: toadflax family, especially common toadflax (*Linaria vulgaris*) & alpine toadflax (*L. alpina*), & *Antirrhinum hispanicum*.

♀ N Spain

Mellicta varia Grisons' Fritillary

Distribution • C Alps. 1500–2500 m. C Apennines. 1200–2600 m. Very sporadic & local.

Description • Ups gc orange with black markings characteristic of Fritillary group; upf pd line often thin & broken; discal mark in s1b variable, often dumbbell- or club-shaped, repeated on unf with conspicuous adjoining black basal bar. Female ups often with extensive greyish suffusion. Most habitats not shared with other members of group.

Flight-period • Late June to late August in one brood. Emergence date depends on altitude.

Habitat • Sheltered flowery slopes & hollows with short grass. LHPs: alpine plantain (*Plantago alpina*), spring gentian (*Gentiana verna*), & trumpet gentian (*G. acaulis*).

Mellicta parthenoides Meadow Fritillary

Distribution • Iberia to NW France, SW Germany (extremely local), W Switzerland, & NW Italy. 400–2400 m. Widespread; locally common in Iberia, very sporadic elsewhere.

Description & variation • Male ups bright fulvous to orange, hw discal field usually clear of black markings; upf marginal & submarginal black lines of uniform thickness; pd line fine, often disrupted in s4; black discal markings heavy; upf discal mark in s1b variable in thickness but distinctly oblique (cf. *M. varia*); uns markings clear & distinctive. Regionally & locally very variable. § Spain. Ups markings generally better defined, contrasting sharply with clear, bright orangey-fulvous gc.

Flight-period • Higher altitudes: early June to July in one brood. Lower altitudes: May–June & August–September in two broods.

Habitat • Open, flowery, grassy, often bushy places in woodland margins. LHPs: ribwort plantain (*Plantago* *lanceolata*), alpine plantain (*P. alpina*), & hoary plantain (*P. media*).

♂ E Spain

♂ E Spain

Mellicta aurelia **Nickerl's Fritillary**

Distribution • W France to Latvia, Balkans, & N Greece
(very local). 100–1500 m. Generally widespread, but local in
western range, extremely sporadic & very local in eastern
range.

Description & variation • Male ups gc orange to fulvous,
sometimes with dusky suffusion; spacing of marginal,
submarginal, pd, & discal black lines roughly even, which, in
company with black veins, create uniform, chequered, fulvous-
orange gc pattern (cf. *M. parthenoides*); unh double marginal
lines infilled yellow, slightly darker than adjacent spots
(cf. *M. britomartis*). Female ups gc generally somewhat paler.

Flight-period • Early June to late July in one brood.
Emergence prolonged in some southern localities.

Habitat • Open, dry or damp, grassy & flowery places,
including heaths & mosses, usually with sparse bushes or small
trees. LHP: ribwort plantain (*Plantago lanceolata*).

Mellicta britomartis **Assmann's Fritillary**

Distribution • N Italy (extremely sporadic & local).
C E Germany to SE Sweden, Lithuania, through Slovenia &
C Balkans to N Bulgaria. 300–900 m. Generally very sporadic
& local, especially near limits of range.

Description • Ups resembles *M. aurelia*, but larger. Male
ups gc darker, lacking dusky suffusion, black markings heavier,
less regular; unh submarginal border often with internal
chocolate brown marks; double marginal lines infilled dusky
orange or brown – distinctly darker tone than adjacent yellow
spots (cf. *M. aurelia*).

Flight-period • Late May to early August, generally in one
brood; possibly in two broods in NW Italy (Ticino Valley).

Habitat • Warm, sheltered, grassy, & bushy places at
woodland margins. LHPs include ribwort plantain (*Plantago
lanceolata*), large speedwell (*Veronica teucrium*), & yellow rattle
(*Rhianthus minor*).

Mellicta asteria Little Fritillary

Distribution • C & E Alps. 2000–2700 m. Extremely sporadic & very local.

Description • Small; ups often with heavy grey suffusion in both sexes. Not easily confused with other species except *Eurodryas aurinia debilis*, which lacks sharply defined & distinctive unh pd orange band.

Flight-period • Early July to late August in one brood.

Habitat • Confined to alpine tundra, open valleys, & slopes with short grass. LHP: alpine plantain (*Plantago alpina*).

Behaviour • Fast & very low in flight, often settling amongst stones or shrubs to avoid strong winds.

Hypodryas maturna Scarce Fritillary

Distribution • E France, through S Germany to S Fennoscandia, Albania, & Bulgaria. 200–1000 m. Extremely sporadic & local.

Flight-period • Late May to early July in one brood.

Description • Ups & uns colouring & pattern of markings very distinctive. Related species do not occur in same habitat.

Habitat • Small, bushy clearings in mature woodland containing young ash or aspen trees, often in damp limestone valleys. LHPs, before hibernation, ash (*Fraxinus excelsior*) or aspen (*Populus tremula*); after hibernation, additional LHPs include ribwort plantain (*Plantago lanceolata*), germander speedwell (*Veronica chamaedrys*), honeysuckle (*Lonicera periclymenum*), & devil's bit scabious (*Succisa pratensis*).

♀ E France

Behaviour • Rests & roosts in trees. In spring & summer, seems to prefer nectar of shrubs, e.g. blackthorn (*Prunus spinosa*), privet (*Lingustrum vulgare*), bramble (*Rubus fruticosus*), & wayfaring tree (*Viburnum lantana*), to that of low-growing herbs.

Conservation • Decline in many regions (N France, S Germany, & S Sweden) attributed to changes in forestry management, land drainage, & habitat destruction for agricultural purposes.

♀ E France

♂ E France

Hypodryas intermedia **Asian Fritillary**

Distribution · C Alps of France, S Switzerland, N Italy to Slovenia (Julian Alps & associated mountains). 1500–2400 m, more often 1700–2000 m. Sporadic & very local.

♂ C Switzerland

Description · Male superficially resembles *H. iduna* & *H. maturna* but does not occur in same region. Female unh pale discal band with medial, thin black line, often broken (cf. *H. cynthia*).

Flight-period · Late June to early August in one brood.

Habitat · Light coniferous woodland clearings or sheltered gullies with low-growing shrubs. LHP: blue-berried honeysuckle (*Lonicera caerulea*).

♂ C Switzerland

Hypodryas cynthia **Cynthia's Fritillary**

Distribution · C Alps of France, Switzerland, Italy, & Austria. 900–3000 m. Bulgaria (Pirin Mts & Rila Mts). 2000–2800 m. Sporadic & local.

Description & variation · Male ups red marginal, submarginal bands & discal spots in striking contrast to white discal & pd areas & heavy (variable) black markings; unh resembles *H. iduna*. Female ups almost unicoloured, dusky orange with well-defined black markings; unh yellow discal band lacking thin black line. § Austria, above 1800 m, f. *alpicola*. Smaller, darker. Male ups with extensive black suffusion, sometimes almost obscuring pale submarginal band; red markings reduced, sometimes absent on fw.

♂ SW Bulgaria

Flight-period · Late June to early August in one brood.

Habitat · Open, grassy slopes with low shrubs, often dominated by juniper (*Juniperous*). LHPs: alpine plantain (*Plantago alpina*) & long-spurred pansy (*Viola calcarata*).

Hypodryas iduna **Lappland Fritillary**

Distribution · Lappland: Arctic Circle to Arctic Sea. 300–700 m. Extremely sporadic; very local but sometimes common.

Description · Superficially resembles *H. intermedia* & *H. maturna* but does not occur in same region.

Flight-period · Late June to mid-July. Date of emergence depends on local weather conditions.

Habitat · Dry, damp, or wet heathland with sparse willow, birch, & wortleberry scrub, often near permanent water. LHPs include alpine speedwell (*Veronica alpina*), rock speedwell

♂ N Sweden

(*V. fructicans*), & bog wortleberry (*Vaccinium uliginosum*).

Behaviour · Males appear surprisingly grey in low, very rapid flight, giving striking impression of skipper butterfly. Often bask or seek shelter from strong winds amongst dwarf birch (*Betula nana*).

Eurodryas aurinia **Marsh Fritillary**

Distribution • Most of Europe, including W British Isles to SE Fennoscandia, Balkans, & N Greece. Absent from Mediterranean islands. 0–2600 m. Generally widespread & local; very sporadic at eastern limits of range.

Description & variation • Probably most variable of all European Fritillaries in respect to size, ups gc, & ups black markings, including dark suffusion. Nonetheless, in overall appearance easily separated from all other species. In general, smaller, darker forms occur in northern range or at higher altitudes in southern range, whilst larger, brighter forms prevalent at low altitude, especially in warmer regions of Mediterranean. These extremes represented by following two forms. § C Alps & Pyrenees, 1500–2600 m, *E. a. debilis*. Small; ups gc yellow; submarginal band & discal spots reddish, normal dark markings extensive. At highest altitudes in Alps, ups pale areas further reduced by dark suffusion (resembles *Mellicta asteria* closely, for which it can easily be mistaken in flight; at rest, dark veins & well-defined orange submarginal band of *M. asteria* enables ready separation). § S Iberia, *E. a. beckeri*. Very large; ups bright brick red, usually with paler discal band & pale marginal chevrons uph; black markings bold. § Ireland, f. *hibernica*. Prominent, red submarginal band & red discal marking contrasting sharply with very pale marginal chevrons, pd band, cellular spots, & heavy, black markings – overall a very distinctive race.

Flight-period • One brood. Emergence date greatly depends on region & altitude. S Portugal: early April to late May. Spain: late May to late June. Alps & Pyrenees: late June to late August. Elsewhere: generally May–July.

Habitat • Adapted to wide range of habitat types. Dry or damp, hot or cool, acidic or alkaline soils in flowery, grassy places, heathland, deciduous or coniferous woodland clearings, or sheltered sites on exposed, alpine slopes. LHPs, at lower altitudes, include devil's-bit scabious (*Succisa pratensis*), honeysuckle (*Lonicera periclymenum*), & small scabious (*Scabiosa columbaria*); at high altitudes in C Alps & Pyrenees: gentians (*Gentiana*) & viscid primrose (*Primula viscosa*).

Behaviour • Characteristic low, gliding flight, & frequent settling to bask with open wings in sheltered, sunny spot, often sufficient to betray identity.

Conservation • Decline in E Britain in 20th century often attributed to interference with, or destruction of habitat. However, climatic change may be an important additional, if not overriding factor. Absence in or disappearance from seemingly suitable habitats, especially in NE Scotland, may be attributable to unreliability of sunshine, a prerequisite for health of species in larval stage.

♀ NE Greece

♂ NW Greece

♀ NE Greece

Eurodryas desfontainii **Spanish Fritillary**

Distribution • S Portugal (Algarve, 100–200 m). SW & E Spain. France (E Pyrenees; very local). 50–1800 m. Sporadic & very local.

♂ C Spain

♂ & ♀ (top) C Spain

Description • Superficially resembles *E. aurinia*. Ups red submarginal band contrasting sharply with pale yellow pd band; ups & uns black marginal lines, veins, & pale marginal chevrons well-defined.

Flight-period • Mid-April to early June in one brood.

Habitat • Hot, dry, grassy scrub, rocky gullies, dry stream beds & neglected areas of cultivation. LHPs include teasel (*Dipsacus fullonum*), *Cephalaria leucantha*, & field scabious (*Knautia arvensis*).

♂ & ♀ (top) C Spain

Satyridae

An extensive family of small, medium, or large butterflies well-represented in Europe. Most species are some shade of brown, the striking exception being the Marbled Whites. Almost invariably, one or more of the wing-surfaces contain distinctive markings known technically as ocelli but often referred to as eye-spots. These are intended, evidently, to divert the point of attack of a predatory bird or lizard from the more vital parts of the insect. The genus *Erebia* is a large and distinctive group of mostly small or medium-sized butterflies, characterized and easily recognized by their dark brown or sometimes almost black ground colour. Characteristically, these are mountain butterflies and it is believed that their dark colour is due to the need to absorb heat directly from the sun in the cooler conditions at higher altitudes. Many are endemic European species, and more than a dozen are confined to the Central Alps. As far as is known, the larvae of all European satyrid butterflies feed on grasses.

Melanargia galathea **Marbled White**

Distribution • N Spain (Cantabrian Mts to Pyrenees), through most of C & S Europe, including S Wales, S England, & Sicily, to Lithuania, Balkans, & C Greece. Records for Peloponnese require confirmation. 0–1750 m. Generally widespread & common.

Description & variation • Upf cell not crossed by black bar (cf. *M. russiae*); cell-base with limited dark greyish suffusion (cf. *M. lachesis*). Ups black suffusion variable, incidence greatest in south-east range, especially in NE Italy, where, in extreme & very striking examples, ups almost entirely black. Ups & uns gc sometimes yellowish, with female upf costa buff – commonly associated with darker forms in south-east range. In all populations, unh sometimes uniformly white, unmarked (commoner in southern range).

Flight-period • Generally June–July in one brood, but records span late May to early September.

Habitat • Grassy, flowery, & bushy places in a wide variety of habitat types. LHPs: a wide range of grasses.

Behaviour • Both sexes are avid feeders on nectar-

rich plants, especially thistles (*Cirsium* & *Carduus*), knapweeds (*Centaureae*), & scabious (*Scabiosa*).

♂ N Greece

♂ N Greece

Melanargia lachesis

Distribution • Portugal & Spain, through S France to Rhône Valley. 0–1600 m. Widespread & common.

Description & variation • Upf cell white, not crossed by black bar (cf. *M. g. galathea* & *M. russiae*). Ups dark suffusion variable in extent & colour (brownish to black); uns gc whitish or yellowish; unh sometimes uniformly white (cf. *M. galathea*).

♂ S Spain

Flight-period • Early June to early August in one brood.

Habitat • Dry, grassy, bushy, flowery places. LHPs: a wide range of grasses.

Melanargia russiae **Esper's Marbled White**

Distribution • N & E Iberia. S France. C & S Italy. N Sicily. Albania. Republic of Macedonia. N C Greece. 600–2100 m. Very sporadic & local, especially in eastern range.

Description & variation • Upf cell crossed with irregular, usually 'jagged', black bar originating at or very close to junction of v2 with cell (distinction from *M. larissa*), often partly obscured by greyish or brownish suffusion (cf. *M. galathea*, *M. lachesis*, & *M. larissa*). Generally, in overall appearance, easily separated from allied species. § South-east range. Ups dark suffusion generally more extensive, but regionally & locally variable.

♂ N Greece

Flight-period • Late June to mid-August in one brood.

Habitat • Dry, grassy, rocky slopes or gullies, often in bushy clearings in sparse woodland. LHPs: grasses, including *Stipa pennata*, *Aegilops geniculata*, tor grass (*Brachypodium pinnatum*), & slender brome (*B. sylvaticum*).

Melanargia larissa **Balkan Marbled White**

Distribution · Coastal regions of Croatia, through S Balkans to European Turkey & Greece, including Corfu, Levkas, Limnos, Lesbos, & Siros. 0–2150 m, usually below 1500 m. Locally common.

♂ N Greece

Description & variation · Upf, distal one-third of cell closed by dark, often irregular line (usually *not* 'jagged') originating *between* junction of v2 & v3 with cell (distinction from *M. russiae*); basal area of cell largely suffused fuscous grey; ups basal suffusion usually extensive but variable (cf. *M. russiae*). Female forms with unh creamy buff or whitish occur in Greece (cf. *M. galathea*).

Flight-period · Late May to early August in one brood.

Habitat · Warm, dry, grassy, flowery places, often amongst bushes & rocks in open woodland. LHPs: grasses.

♀ S Greece

♀ (top) & ♂ N Greece

♂ C Greece

Melanargia occitanica
Western Marbled White

Distribution • Portugal through S France to NW Italy. 0–1500 m. Widespread but local. NW Sicily. 600–1000 m. Very local.

Description & variation • Unh veins brownish – overall pattern very distinctive. § Sicily, *M. o. pherusa*. Ups black markings reduced; uph & unh ocelli smaller, sometimes absent; unh veins brown, finer.

Flight-period • Late April to late June in one brood.

Habitat • Hot, dry, grassy, rocky places. LHPs include tor grass (*Brachypodium pinnatum*) & cock's-foot grass (*Dactylis glomerata*).

♀ S Spain

♂ S Spain

Melanargia arge Italian Marbled White

Distribution • S Italy. 350–1500 m. Local but common.

Description & variation • Resembles *M. occitanica pherusa*. Ups & uns black markings reduced; black mark at cell-end often enclosing bluish scales; base of s3 usually with some dark suffusion; unh veins black.

Flight-period • Early May to mid-June in one brood.

Habitat • Dry, rocky, & grassy places. LHPs: grasses.

♀ S Spain

Melanargia ines Spanish Marbled White

Distribution · Portugal. Spain (south of Cantabrian Mts & Pyrenees). 50–1500 m. Widespread, locally common.

Description · Superficially similar to *M. arge* & *M. occitanica* but without unh dark, longitudinal line in s1b; ups & uns quite distinctive in overall appearance.

♂ S Spain

Flight-period · Late March to late June in one brood.

Habitat & behaviour · Dry, grassy, & rocky places. LHPs include tor grass (*Brachypodium pinnatum*). Habitats often shared with *M. occitanica.*

♀ S Spain

Hipparchia fagi Woodland Grayling

Distribution • N Spain through C Germany to S Greece. Absent from Mediterranean islands except Sicily & Levkas. 50–1800 m, usually below 1000 m. Local.

Description • Resembles *H. alcyone* & *H. syriaca* very closely. Reliable separation probably impossible without reference to male/female genitalia.

Flight-period • Early June to mid-September in one brood.

Habitat • Bushy, grassy, woodland clearings; margins of pine forests. LHPs: grasses, including upright brome (*Bromus erectus*) & red fescue (*Festuca rubra*).

Behaviour • Often rests in shade on tree-trunks or interior of bushes.

♂ N Greece   ♀ N Greece

Hipparchia alcyone Rock Grayling

Distribution • S Spain to C E Europe, S Norway, & SE Latvia. 0–1600 m, usually above 500 m. Local & generally scarce in northern range. Distribution uncertain in many regions owing to possible confusion with *H. fagi*.

Description • Resembles *H. fagi* & *H. syriaca* very closely. Reliable separation probably impossible without reference to male/female genitalia.

Flight-period • Late June to mid-August in one brood.

♂ E Spain

Habitat • Bushy, grassy, rocky, woodland clearings or margins, more commonly associated with pinewoods. LHPs: tor grass (*Brachypodium pinnatum*), slender brome (*B. sylvaticum*), & sheep's fescue (*Festuca ovina*).

Hipparchia syriaca **Eastern Rock Grayling**

Distribution • S Balkans & Greece, including Corfu, Kefalonia, Thassos, Lesbos, Chios, Samos, & Rhodes. 0–1300 m. Sporadic & local.

Description & variation • Resembles *H. fagi* very closely. Reliable separation very difficult without examination of genitalia. Male ups pale pd band, often partly or largely obscured by dark suffusion.

♂ Samos, Greece

Flight-period • Generally June–August in one brood, but records span early May to mid-September.

Habitat • Hot, dry, bushy places in woodland, usually open pinewoods.

Behaviour • Often rests in shade on tree-trunks. Several males may assemble on one tree.

♂ Samos, Greece

♀ Samos, Greece

Hipparchia neomiris **Corsican Grayling**

Distribution • Known only from Corsica, Sardinia, Capraia, & Elba. 300–2000 m. Locally common.

Description • Ups brown, with wide, yellowy-orange pd bands; unh white pd band usually bold. Distinctive & easily distinguished from *H. aristaeus*, with which it sometimes occurs.

Flight-period • Mid-June to August in one brood.

Habitat • Rocky slopes with low-growing herbs & scrub, usually in or near open pinewoods. LHPs: grasses.

♂ Corsica, France

Hipparchia volgensis **Delattin's Grayling**

Distribution • Albania. SW Serbia (Kosovo). Republic of Macedonia. Bulgaria. Greece, including Zakynthos. 700–1700 m. Locally common.

Description • Resembles *H. semele* very closely, but not known to occupy same localities. Examination of genitalia required for reliable separation.

Flight-period • Early June to August in one brood.

Habitat • Rocky places, including screes, usually in open, dry pinewoods or scrub. LHPs: grasses.

♂ N Greece

♀ N Greece

Hipparchia semele Grayling

Distribution • Most of Europe, including Sicily
(1000–1800 m), Lipari Islands (0–500 m) through British Isles
& S Fennoscandia to Albania & Bulgaria. 0–2000 m.
Distribution in E & SE Europe uncertain due to confusion
with closely similar species. Generally widespread &
common.

Description & variation • Male ups brown; upf with
prominent sex-brand; upf yellowish or orange pd band
extremely variable, sometimes reduced to obscure marks
enclosing dark, white-pupilled ocelli in s3 & s5; uph pd better
developed, broken by veins, with ocellus in s2; unf disc
yellowish, wing margins & costa greyish, somewhat mottled;
white-pupilled, dark ocelli in s3 & s5 prominent; unh grey,
finely speckled with darker & paler tones; fine, black
mediobasal line very irregular, similar discal line sharply
defining limit of irregular & somewhat diffuse white pd band
(variable); ocellus in s2 often obscure. Female larger; all ups
markings better developed; uph pd band usually darker
(often orange) than that of upf; unh white pd band more
diffuse. § Lipari Islands, *H. s. leighibi.* Larger; ups bright
orange markings extensive, sometimes extending to wing-
bases. (Very similar to *H. aristaeus blachieri,* from which
probably indistinguishable without examination of genitalia.
However, two forms not known to occur together.) Marked
variation in size, colour, & markings, especially unh, seems due
largely to local (ecological) conditions. Darker forms tend to
occur on darker, usually acidic soils; paler forms are prevalent
on lighter, alkaline soils, usually limestone.

Flight-period • May–September in one brood. Emergence
date much dependent upon locality.

Habitat • Well-adapted to diverse range of habitat types,
including dry chalk or limestone grassland, wet or dry
heathland, sand-dunes, rocky slopes, coastal cliffs, open
woodland, or scrub. LHPs: wide variety of grasses.

Behaviour • When settling to rest or
roost, fw apex always left projecting slightly
beyond hw for several seconds, thus
exposing subapical ocellus (eye-spot).
Thereafter, fw is dropped behind well-
camouflaged hw, where it remains until
insect takes flight or is disturbed by sudden
movement in its field of vision. Evidently,
this behaviour is designed to deflect attack-
point of predatory birds or lizards, which
may have observed insect settling, away from
more vulnerable head, thorax, & abdomen.
A very brief but very critical period is thus
provided, usually allowing butterfly to evade
its predators before serious damage is
inflicted on vital body-parts. Same
behaviour is displayed by other grayling
butterflies & many other similarly marked
species.

Hipparchia cretica **Cretan Grayling**

Distribution • Crete. 100–1500 m. Widespread, locally common.

Description • Resembles *H. semele* very closely, but no related species occurs on Crete.

Flight-period • Mid-May to mid-August in one brood.

Habitat • Dry, rocky ground with bushes & sparse trees; common in olive groves.

Behaviour • Rests on shaded side of tree-trunks in very hot conditions.

♂ Crete, Greece

Hipparchia christenseni **Karpathian Grayling**

Distribution • Karpathos (Mt Lastros & Kali Limnos). 300–750 m. Very local.

Description • Resembles *H. semele* & *H. cretica* very closely, but no related species occurs on Karpathos. Examination of genitalia required for reliable separation.

Flight-period • Early to late June in one brood.

Habitat • Dry, bushy, stony places & pinewood clearings.

Behaviour • Often rests & roosts on trunks of pine trees. In treeless, bushy terrain, males usually rest on stones.

Hipparchia aristaeus **Southern Grayling**

Distribution • Madeira (800–1800 m). Corsica. Sardinia. Tyrrhenian islands of Capraia, Elba, Giglio, & Ponza (500–1800 m). Sicily (700–1900 m). S Balkans & Greece, including Levkas, Thassos, & several Aegean islands, but excluding Crete, Karpathos, & Rhodes (50–1600 m). Local, sometimes uncommon.

Description & variation • Regionally extremely variable. § Corsica, Sardinia, & associated islands, *H. a. aristaeus*. Superficially similar to *H. semele* in basic characters; larger; ups orange markings brighter, more extensive, especially in female where orange markings may extend to upf & uph wing-bases; uns markings closely similar. Not known to occupy same localities as related species except *H. neomiris*. § Madeira, *H. a. maderensis*. Smaller. Male ups markings largely obscured by dark brown suffusion; unf discal area dusky yellow; unh white discal band usually prominent. Female ups markings better developed. § Sicily, *H. a. blachieri*. Resembles nominate form, but larger; ups orange areas brighter, more extensive. Closely resembles *H. semele leighibi* but not known to occur in same localities. § S Balkans & Greece, *H. a. senthes*. Ups yellowish areas reduced, suffused greyish brown. Resembles *H. volgensis* closely; not reliably separable without examination of genitalia.

Flight-period • Generally May to late August in one brood. Emergence date depends on locality. Madeira: late July to September.

Habitat • Rocky slopes or gullies, amongst scrub or sparse woodland. LHPs: grasses.

♂ *H. a. senthes*, N Greece

Hipparchia azorina **Azores' Grayling**

Distribution • Azores. Restricted to Pico, Sao Jorge, Faial, & Terceira). 480–2000 m. Local, generally uncommon.

Description • Male ups gc, dusky greyish brown, dull yellowish pd & discal markings obscure; upf sex-brand conspicuous; unh gc very dark, contrasting prominent, irregular white pd band. Female ups gc lighter, diffuse yellow markings more distinct.

Flight-period • June–October in one brood.

Habitat • Sheltered, grassy slopes. LHP: fescue (*Festuca jubata*).

Behaviour • Both sexes attracted to nectar of herbs & low-growing shrubs.

Hipparchia caldeirense **Oehmig's Grayling**

Distribution • Azores. Restricted to Flores (Caldeira Seca & Pico dos Sete Pes). Above 700 m. Local, generally uncommon.

Description • Resembles *H. azorina*, but ups yellowish markings better developed. Male upf without sex-brand.

Flight-period • June to late September in one brood.

Habitat • Sheltered hollows, gullies, & small valleys on grassy slopes. LHP: fescue (*Festuca jubata*).

Behaviour • As for *H. azorina*.

Hipparchia miguelensis **Le Cerf's Grayling**

Distribution • Azores. Restricted to Sao Miguel. 600–1000 m. Generally uncommon.

Description • Otherwise similar to *H. caldeirense*. Male upf sex-brand conspicuous.

Flight-period • Late June to late September in one brood.

Habitat • Sheltered, grassy places. LHP: fescue (*Festuca jubata*).

Behaviour • Both sexes attracted to nectar of herbs & low-growing shrubs, including ling (*Calluna vulgaris*).

Hipparchia mersina

Distribution • Greece: in Europe, known only from Lesbos & Samos. 150–1150 m. Locally common.

Description • Resembles *H. pellucida*, with which it may be easily confused. No similar member of genus occurs in same locality.

Flight-period • Mid-May to mid-July in one brood.

Habitat • Dry, grassy clearings amongst rocks, sparse bushes, & trees. LHPs: grasses.

♂ Samos, Greece

Hipparchia pellucida

Distribution • Greece: in Europe, known only from Lesbos & Ikaria. 400–700 m. Locally common.

Description • Resembles *H. mersina*, with which it may be easily confused. No similar member of genus occurs in same locality.

Flight-period • Late May to July in one brood (data limited).

Habitat • Dry, grassy, or stony ground with sparse vegetation including bushes & pine trees. LHPs: grasses.

Neohipparchia statilinus Tree Grayling

Distribution • Most of S Europe, including Mediterranean islands of Sicily, Elba, Corfu, & Thassos, through NW France to S Switzerland, E Poland, Lithuania, & Greece. 0–1400 m. Widespread; locally common in Mediterranean region, very sporadic & local in northern range.

Description & variation • Unh submarginal line variable, often obscure (cf. *N. fatua*). Regionally & locally very variable, especially unh gc & markings: dark, greyish-brown to pale grey, with variable brownish tint & greyish-brown mottling; variation in development of submarginal, discal, mediobasal lines varies independently of gc. Generally easily identified, despite variation.

Flight-period • Late June to October in one brood, with peak emergence generally late July to early August in Mediterranean region.

Habitat • Hot, dry, rocky places amongst scrub or scattered trees. LHPs: grasses, including upright brome (*Bromus erectus*), barren brome (*B. sterilis*), *Bothriochola ischaemum*, *Stipa pennata*, & *Lygaeum spartum*.

Behaviour • Usually rests on stones or soil. In hottest conditions, may seek shade of tree-trunks, where it also sometimes roosts. Not greatly attracted to flowers, but females sometimes take nectar from thyme (*Thymus*).

♂ N Greece

Neohipparchia fatua **Freyer's Grayling**

Distribution • SW & S Balkans & Greece, including Lefkas, Kithira, Thassos, Paros, Spetses, & most E Aegean islands (not reported from Corfu or Crete). 0–600 m. Sporadic, locally common.

Description • Resembles *N. statilinus*, but larger. Male ups very dark; unh mediobasal, discal, & submarginal fine, very dark lines well-defined (cf. *N. statilinus*).

Flight-period • Late May to early October in one brood according to locality.

♀ S Greece

Habitat • Hot, dry, rocky, or stony places amongst scrub or pine trees; also olive groves & orchards. LHP: grasses.

Behaviour • Seeks shade of tree-trunks in hottest part of day; generally flies very little unless disturbed.

Pseudotergumia fidia **Striped Grayling**

Distribution • Most of Iberia, including Mallorca, through S France to NW Italy (Maritime Alps). 0–1400 m. Widespread, very local.

Description • Ups gc dark, greyish brown; upf white-pupilled ocelli in s2 & s5, & white spots in s3 & s4, very distinctive; unh colour & colour pattern enables easy separation from other species.

Flight-period • Late June to late August in one brood.

♀ S Spain

Habitat • Hot, dry, rocky slopes, or dry ground with sparse, low-growing vegetation, often amongst bushes or scattered trees. LHPs: grasses, including *Cynodon dactylon*, cock's-foot (*Dactylis glomerata*), & *Milium multiflorum*.

Pseudotergumia wyssii **Canary Grayling**

Distribution • Known only from Canary Islands: Tenerife (1400–2000 m), N & W Gomera (200–400 m), NE La Palma (400–1300 m), Hierro (300–1500 m), & Gran Canaria (400–2000 m). Generally very local.

Description & variation • Ups resembles *P. fidia*, but markings less prominent in both sexes; unh gc paler, yellowish brown, markings more obscure. No related species occurs in same habitats. Minor differences within & between populations of different islands may be apparent upon close examination.

Flight-period • April to early September in one brood. Emergence date depends on locality & altitude.

Habitat • Rocky gullies & slopes in scrub margins of laurel or pine forests. LHP(s): grasses.

Chazara briseis **Hermit**

Distribution • Spain, through C & S France, Italy, Sicily, C & S Balkans, to S Greece. 0–2000 m, more generally below 1600 m. Sporadic in many regions, generally very local.

Description & variation • Male ups gc dark khaki brown with prominent creamy-white pd bands, broken on fw by veins; dark subapical ocellus prominent; unh brown discal band broken by cell; light brownish pd band, separated from darker submarginal band by irregular dark brown line; discal band pale cream or white. Female similar; ups pd band wider, paler, with additional white-pupilled ocellus in s3; unh paler, yellowish grey, finely speckled with black & grey scales. § S Europe. Female ups white markings sometimes replaced by yellowish buff or creamy brown. § SE Europe. Female unh sometimes suffused pink.

Flight-period • Late May to October in one brood. Emergence date depends on locality & altitude.

Habitat • Dry, often hot, stony or grassy places amongst scrub. LHPs: grasses, including upright brome (*Bromus erectus*) & sheep's fescue (*Festuca ovina*).

Behaviour • Males often rest on stones or bare soil.

♂ E Spain   ♀ NW Greece

Chazara prieuri **Southern Hermit**

Distribution • Spain (provinces of Granada, Murcia, Teruel, Madrid, Huesca, Zaragoza, Alicante, & Mallorca). 850–1450 m. Very sporadic, very local, & often very scarce.

Description & variation • Male upf with conspicuous buff patch in cell. Female ups white markings replaced by orangey brown in about 50% of specimens. Both sexes: unh veins conspicuously pale (cf. *C. briseis*).

Flight-period • Mid-July to mid-August in one brood.

♂ E Spain

Habitat • Hot, rocky gullies, or dry, grassy, & stony slopes with sparse pine-trees or scrub. LHP: albardine (*Lygeum spartum*).

Behaviour • Both sexes often rest on bare, stony ground amongst stones. Very wary, but not easily disturbed from well-concealed resting sites.

Pseudochazara graeca **Grecian Grayling**

Distribution • Republic of Macedonia (Pelister massif). Greece: most mountains from Smolikas & Olympus massifs to Mt Taygetos in S Peloponnese. 1000–2200 m. Very local, often very common.

Description & variation • Ups gc greyish brown; upf yellowish, well-marked pd bands broken at v4; uph pd band wider, orange or yellowy orange; upf dark ocelli in s2 & s5 prominent, uph ocellus in s2 very small, sometimes absent; unh greyish, finely speckled with paler & darker tones; dusky orange submarginal band internally bordered by paler,

♂ NW Greece

yellowish area. Overall character, especially unh coloration, enables easy separation from similar species occurring in same habitat. Small but systematic variation in size & coloration, especially unh, evident in almost every population. Unh coloration, particularly, appears to correspond to colour of rocks & soil in habitat in which butterfly must rest, roost, & seek refuge from predatory birds & lizards. § Smolikas massif, Timfi Mts, & Katarapass (1200–1600 m). Ups suffused smoky brown; upf ocelli obscure, often lacking white pupils; unh browner, darker.

Flight-period • Mid-July to late August in one brood.

Habitat • Mostly open, grassy places amongst limestone rocks above tree-line; occasionally on dry, grassy slopes amongst scrub & sparse pine trees. LHPs: grasses, including sweet vernal grass (*Anthoxanthum odoratum*).

Pseudochazara hippolyte **Nevada Grayling**

Distribution • S Spain: Sierra Nevada (2000–2700 m), Sierra de los Filabres (1850–2020 m), Sierra de Gádor (2000–2200 m), Sierra de Maria (1400–2040 m), & Sierra de Guillimona (1500–2000 m). Extremely local, often very common.

Description & variation • Ups gc greyish brown; yellow submarginal borders complete, crossed by dark veins; upf dark ocelli in s2 & s5 prominent; uph ocellus in s2 variable, usually very small; unh greyish brown, with well-defined irregular basal, discal, & pd dark lines; pd band paler, finely speckled with dark scales. Minor variation in size & gc apparent for each isolated mountain population.

♀ S Spain

Flight-period • Late June to late July in one brood.

Habitat • Dry, grassy, stony slopes, often with large areas of barren, stony soil. LHP: sheep's fescue (*Festuca ovina*).

Pseudochazara geyeri Grey Asian Grayling

Distribution • Albania (mountains bordering Lake Ohrid). SW Republic of Macedonia (mountains bordering Lake Prespa, including Pelister massif). NW Greece (Mt Malimadi & Triklarion Mts). 1450–1700 m. Extremely sporadic, locally common.

♂ NW Greece

♀ NW Greece

Description • Ups gc pale, greyish yellow-brown; ups pd bands dull, greyish yellow; uph undulating outer margin of pd band well-marked & characteristic; upf ocelli in s2 & s5 well-marked; unh dark ocellus in s2 very small; unh pattern of jagged submarginal, discal dark lines & white pd band distinctive. Easily distinguished from similar species occupying same habitat.

Flight-period • Mid-July to late August in one brood.

Habitat • Dry, rocky, & grassy limestone slopes above tree-line. LHP(s): grasses.

Behaviour • Often settles on stones or bare soil, & easily disturbed. Both sexes strongly attracted to nectar of knapweed (*Centaureae*).

Pseudochazara mamurra Brown's Grayling

Distribution • NW Greece (known only from district of Ioannina). 650 m. Extremely local & very uncommon.

Description • Resembles *P. graeca*. Ups gc paler; pd bands pale orange, wider; unh paler with light sandy brown overtones.

Flight-period & habitat • Information very limited. Recorded in July on rocky ground with reddish soil.

Pseudochazara orestes Dils' Grayling

Distribution • Known only from N Greece (Phalakron massif, Menikion Mts, & Mt Orvilos) & Bulgaria (Pirin Mts). 800–1800 m. Very local, sometimes common.

Description • Ups medium brown; pd bright orange bands sharply defined, broken only by brown veins; upf white-pupilled, dark ocelli in s2 & s5 bold, white spots in s3 & s4 usually prominent; unh marginal & basal areas greyish with darker speckling; submarginal band similar with brown tone; pd band white.

♂ N Greece

Flight-period • Mid-June to late July in one brood.

Habitat • Hot, dry, mostly south-facing, rocky or stony slopes, sometimes with small trees & shrubs. LHPs: grasses.

Behaviour • In very hot conditions, both sexes retire to shade of rock crevices & ledges.

Pseudochazara mniszechii **Dark Grayling**

Distribution • NW Greece (Vernon Mts, Grammos Mt, & Mt Smolikas). 850–1500 m. Extremely sporadic & local, but often common.

Description • Ups medium brown; pd bands rich orange, sharply defined, broken by brown veins; upf white-pupilled, dark ocelli in s2 & s5 bold, white oval spots in s3 & s4 usually prominent, especially in female. Male unh yellowish brown gc dark marginal & pale pd bands largely obscured by dark speckling. Female unh similar, but darker, richer brown.

Flight-period • Mid-July to late August in one brood.

♀ NW Greece

Habitat • Dry slopes with sparse grasses & herbs amongst light scrub & scattered trees. LHPs: grasses.

Behaviour • Rests on bare patches of soil, rather than stones. Roosts in well-concealed spaces, between or under large stones. Both sexes strongly attracted to nectar of a purple thistle.

Pseudochazara cingovskii
Macedonian Grayling

Distribution • Republic of Macedonia (known only from Pletvar massif). 1000–1200 m. Extremely local, usually very common.

Description • Ups greyish brown; pd bands dirty yellow, sharply defined, broken by dark veins; upf white-pupilled, dark ocelli in s2 & s5 bold, white spots in s3 & s4 well-defined. Male unh greyish yellow, speckled dark grey; dark submarginal, discal, & basal lines well-defined; pale pd bands with darker speckling. Female unh similar, paler, all markings better defined.

Flight-period • Late July to early August in one brood.

Habitat • Dry, white limestone rocks with sparse vegetation. LHPs: grasses.

Pseudochazara anthelea
White-banded Grayling

Distribution • S Balkans & Greece, including Lesbos, Samos, Kos, Kalimnos, Chios, Rhodes, & Crete. 500–1800 m. Widespread; very local but often common.

♂ *P. a. amalthea*, NW Greece

♂ *P. a. amalthea*, S Greece

Description & variation • Male ups gc dark brown; upf white pd band, externally bordered orange on hw; upf sex-brand well-defined; upf prominent ocelli in s2 & s5 with small white pupils; unh irregular, white pd band contrasting with dark, mottled gc. Female ups gc medium brown, contrasting with orange pd areas, extended to cell upf; unh orangey brown gc largely obscured by dark speckling; white pd band diffuse. § Mainland Europe & Crete, *P. a. amalthea*. Male superficially almost identical. Female ups pd orange areas replaced by cream & orangey brown; upf orange discal area replaced by medium brown, ocelli large, usually lacking white pupils; unh similar to nominate form. § Rhodes (Mt Ataviros; 500 m). Female ups transitional to nominate form.

Flight-period • Late May to early July in one brood.

Habitat • Dry, stony slopes & gullies, often amongst sparse bushes or trees. LHPs: grasses.

Behaviour • Males territorially very defensive, often 'perching' on rocks to which they quickly return if disturbed.

Oeneis norna **Norse Grayling**

Distribution • C Scandinavia to North Cape & NW Finland. 250–800 m. Widespread, local, sometimes common.

Description • Ups dull, yellowish orange. Extremely variable in size & all wing-markings. Small, lightly marked specimens sometimes resemble *O. bore*, but ups never quite as grey.

♂ N Sweden

Flight-period • Mid-June to mid-July in one brood.

Habitat • Damp, heathy, grassy, or boggy places, usually with birch or willow scrub; sometimes large forest-clearings. LHPs: grasses, including common cat's-tail (*Phleum pratense*), alpine meadow grass (*Poa alpina*), & matgrass (*Nardus stricta*).

♀ N Norway

Oeneis bore **Arctic Grayling**

Distribution • Lappland (67°N to Arctic Sea). 100–1000 m. Sporadic & very local.

Description • Ups gc with distinct greyish tone, lacking 'warmer', brownish tone of *O. norna*; upf & unf with small, round, very pale pd dot in s5, sometimes very obscure but usually present – absent in *O. norna*. Both sexes lose ups scales readily, giving a worn, shiny appearance.

Flight-period • Mid-June to late July in one brood according to season.

Habitat & behaviour • Damp, grassy, & boggy places, characteristically dominated by small, elevated areas of flat, almost barren rock, which appear to be meeting-places for males & females. Males often rest on exposed rock-faces for prolonged periods in warm, sunny, & often windy conditions. LHP: sheep's fescue (*Festuca ovina*).

♂ N Norway   ♀ N Norway

Oeneis glacialis **Alpine Grayling**

Distribution • C Alps. 1400–2900 m. Sporadic, local, & generally not common.

Description • Resembles *O. norna* closely, but unh veins white. Geographically well-separated.

Flight-period • Early June to mid-August in one brood. Emergence date depends on altitude.

Habitat • Dry, grassy places amongst rocks or scree. LHP: sheep's fescue (*Festuca ovina*).

Oeneis jutta **Baltic Grayling**

Distribution • NE Poland to 65°N in Fennoscandia. 100–500 m. Sporadic & very local.

Description • Ups medium greyish brown; ups yellowish orange pd bands & dark ocelli distinctive; unh greyish, with darker speckling; whitish discal & pd areas obscure. Not easily confused with any other species.

♂ C Sweden

♂ C Sweden

Flight-period • Early June to mid-July in one brood. Emergence date depends on latitude & weather conditions.

Habitat • Boggy ground dominated by grasses, usually near open areas of water, always with scattered pine trees, & often bordered by pinewoods.

Behaviour • Usually rests on trunks of pine trees. LHPs: grasses.

Satyrus actaea **Black Satyr**

Distribution • Iberia, through S France to NW Italy (Maritime & Cottian Alps). 100–2000 m. Widespread, locally common.

Description & variation • Ups dark brown to nearly black to pale brown, especially in female; upf subapical, white-pupilled ocellus better developed in female; male upf with androconial patch in s1–3 (cf. *S. ferula*); uns paler pd & discal areas better marked in female, very variable, often obscure. All wing-markings, gc, & size locally, individually, & regionally variable.

Flight-period • Early June to late August in one brood

Habitat • Dry, grassy, rocky places, often scrub or sparse trees. LHPs: grasses.

Behaviour • In hot conditions, both sexes seek shade of rocks. Both sexes strongly attracted to nectar of thistles (*Cirsium* & *Carduus*) & thyme (*Thymus*).

♂ S Spain

Satyrus ferula **Great Sooty Satyr**

Distribution • Spain (extremely local in Pyrenees; 500–1400 m). France (Pyrenees, to Dordogne & Massif Central), through S Switzerland & Italy to S Balkans & Greece. 0–1800 m. Very sporadic & local, especially in western range.

♂ S Switzerland

♀ N Greece

Description • Resembles *S. actaea*. Male ups dark brown with white-pupilled black ocelli in s2 & s5; upf without sex-brand (cf. *S. actaea*); unh small ocellus in s2 often absent; unh dark brown, with pale submarginal & pd bands. Female gc medium brown, with dusky orange pd bands & prominent, white-pupilled, black ocelli in s2 & s5; white spots in s3 & s4; unh gc buff or pale yellowish brown with pale submarginal & pd bands. § Limestone localities. Female unh brightly marbled white to grey.

Flight-period • Mid-June to early August in one brood.

Habitat • Open, grassy, & rocky slopes, or grassy & bushy woodland clearings. LHP: sheep's fescue (*Festuca ovina*).

♀ N Greece

Minois dryas **Dryad**

Distribution • N Spain, through C & S France to C & S Balkans & C N Greece. 100–1600 m. Sporadic & local, especially near limits of range.

Description • Ups dark, somewhat greyish brown; upf blue-pupilled ocelli in s2 & s5 distinctive, usually large, especially in female, Male unh brown, marginal paler, submarginal band darker; white-pupilled, dark ocellus in s3 usually obscure. Female similar; ups gc paler; upf ocelli larger; unh gc paler, well-marked with darker submarginal band & whitish pd band.

♂ N Greece

Flight-period • Late June to early September in one brood.

Habitat • Grassy, bushy, woodland margins, often humid with lush vegetation, including ferns. LHPs: grasses.

Kanetisa circe **Great Banded Grayling**

Distribution • Most of S Europe through C France & S Poland to E Balkans & Greece, including Corsica, Sardinia, Sicily, Thassos, & Lesbos. 0–1650 m. Widespread & common.

Description • Ups gc very dark brown, boldly marked with white pd bands broken by veins; upf subapical ocellus very dark; unh dark greyish brown with white pd & post-basal bands.

♂ N Greece

Flight-period • Early June to mid-September in one brood.

Habitat • Dry, grassy, bushy places; cultivated ground. LHPs include sheep's fescue (*Festuca ovina*) & upright brome (*Bromus erectus*).

Arethusana arethusa **False Grayling**

Distribution • Iberia & France, through Slovakia to S Balkans & Greece. 0–2000 m. Generally sporadic & local, sometimes common. Very sporadic in C Europe & NW Balkans.

Description & variation • Ups gc dark brown with yellowish tone; ups dusky orange pd band variable, sometimes very obscure, always broken by dark veins; subapical ocellus dark, sometimes with small white pupil, especially in female; unh greyish brown, speckled dark grey or black; white pd band

♂ C Greece

narrow, veins sometimes conspicuously lined white. Regionally, locally, & individually variable.

Flight-period • Mid-July to mid-September in one brood.

Habitat • Grassy, bushy places, woodland borders, or rocky gullies. LHPs: grasses.

Erebia ligea **Arran Brown**

Distribution • S France (Massif Central to Maritime Alps & Vosges Mts) to Czech Republic, S Poland, Balkans, & N Greece. Fennoscandia & Baltic countries. 0–1800 m. Generally widespread, local, usually common.

Description & variation • Male upf with sex-brand (cf. *E. euryale*). Ups reddish pd bands variable in colour & width; ocelli variable in size & number, sometimes small, lacking white pupils.

Flight-period • Mid-July to late August in one brood.

♂ N Norway

Habitat & behaviour • Sheltered, grassy, flowery places in woodland clearings, often damp & humid, sometimes containing an abundance of bracken upon which adults often bask with open wings in sunny or warm, overcast conditions. LHPs include wood sedge (*Carex sylvatica*) & loose-spiked sedge (*C. strigosa*).

♀ N Greece

Erebia euryale Large Ringlet

Distribution • N Spain (Cantabrian Mts & Pyrenees) & S France, through C Alps & C Italy to Carpathian Mts, S Balkans, & N Greece. 750–2500 m, more generally 1200–2000 m. Generally widespread, locally common, sporadic, & local near limits of range.

♂ SE France

Description & variation • Male upf without sex-brand (cf. *E. ligea*). Female unh whitish or yellowish pd band broad, conspicuous. Development of ups pd orangey-red markings & ocelli (number & size) regionally extremely variable.

Flight-period • Late July to August in one brood.

Habitat • Grassy, flowery places in pinewood clearings; sheltered, grassy slopes above tree-line. LHPs include *Sesleria varia*, sheep's fescue (*Festuca ovina*), red fescue (*F. rubra*), wood meadow grass (*Poa nemoralis*), & chalk sedge (*Carex flacca*).

Behaviour • In warm, sometimes sunless, conditions, often basks with open wings in sheltered spots.

♀ SE France

Erebia eriphyle **Eriphyle Ringlet**

Distribution • C Alps. 1200–2250 m. Widespread, very local.

Description • Ups gc dark brown; ups reddish pd bands broken by veins (enclosing 2–4 black points in s2–5) reduced to red spots (variable) on uph but elongate spot in s4 constant; unf often with reddish suffusion extending from border of paler pd band to wing-base; unh reddish orange spots in s2–5 variable, lacking black pupils, spot in s4 larger, always present. (cf. *E. manto* f. *pyrrhula* & *E. melampus*).

Flight-period • Late June to mid-August in one brood.

Habitat • Damp, sheltered, herb-rich, alpine meadows or rocky gullies, often near pinewoods or in open areas amongst scrub. LHPs: sweet vernal grass (*Anthoxanthum odoratum*) & tufted hair-grass (*Deschampsia caespitosa*).

Erebia manto **Yellow-spotted Ringlet**

Distribution • N & E Spain (Cantabrian Mts & Pyrenees), through S France (Massif Central & Vosges Mts), C Alps, & Julian Alps, to Bosnia-Herzegovina & Carpathian Mts. 900–2500 m. Sporadic, locally common.

♂ E France

♀ E France

Description & variation • Size, gc, & markings regionally variable. § Spain & Massif Central. Male ups & uns very dark, almost black; usual uns markings absent or vestigial. Female paler; unh markings often obscure. § Above 1800 m in C Alps, f. *pyrrhula*. Very small; dark; all markings greatly reduced, sometimes absent. § Vosges Mts. Large. Male upf pd reddish-orange band narrow but unbroken; unh pd pale yellowish bands well-developed. Female similar; all markings better developed; unh yellow pd markings sometimes replaced by pure white (a recurrent form in some localities in Switzerland).

Flight-period • Early July to early September in one brood.

Habitat • Damp, grassy, flowery meadows & slopes, often in woodland. LHP: red fescue (*Festuca rubra*).

Erebia claudina **White Speck Ringlet**

Distribution • Known only from Alps of Austria (Hohe Tauern, Salzburg Alps, Niedere Tauern, Seetal Alps, & Sau Alp). 1800–2300 m. Very local.

Description • Male ups & uns gc dark brown; upf reddish pd band broken by veins, enclosing obscure, twin black points in s5 & s6; uph with 2–5 white pd points, 6 on unh; unf reddish discal & paler pd areas obscure. Female similar; all markings better developed; unh gc, dusky yellowish, veins prominent. Distinctive, easily distinguished from other members of genus.

Flight-period • July to early August in one brood.

Habitat • Grassy slopes, usually above tree-line. LHPs: grasses.

♀ Austria

Erebia flavofasciata **Yellow-banded Ringlet**

Distribution • Known only from Alps of C N Italy & SE Switzerland. 1800–2600 m. Very local, usually common.

Description • Ups medium brown; black-pupilled, orange, elongate pd spots somewhat variable in size & number; unh gc brown, black-pupilled, yellow pd spots, usually in complete series, forming band broken only by dark veins. Very distinctive – quite unlike any other member of genus.

Flight-period • Late June to mid-August in one brood.

Habitat • Open grassy slopes. LHP: sheep's fescue (*Festuca ovina*).

Erebia epiphron **Mountain Ringlet**

Distribution • Most mountain ranges of Europe, including Cumbrian Mts (NW England) & Grampian Mts (C Scotland), but excluding C & S Spain, Fennoscandia, C & S Greece, & Mediterranean islands. 350–2700 m. Very sporadic near limits of range, locally often common.

Description & variation • § Nominate form (*E. e. epiphron*), Harz Mts (believed to be extinct). Male upf pd rust-coloured band broad, usually complete, sometimes broken into discrete spots, enclosing black ocelli in s2–5; uph similar pd band enclosing black ocelli in s2–4; unh with red-ringed ocelli in s2–4/5. Female markings better developed; unh reddish pd spots often forming a distinct band; ups & unh black ocelli sometimes enclosing minute white points. Regionally & locally, extremely variable. § England & Scotland, *E. e. mnemon*. All markings reduced, but variable; pd bands darker, often broken into discrete spots, especially unh. § Vosges Mts, *E. e. mackeri*. Ups & upf orangey-red pd band & black ocelli usually well-developed; unh pd markings obscure. § E Pyrenees, *E. e. fauveaui*. Resembles *E. e. mackeri*. Slightly larger; all markings slightly reduced but individually & locally variable. Female unh pd black ocelli ringed orange. § Silesian Mts, *E. e. silesiana*. All markings well-developed; resembles nominate form. § C Alps & W Balkans, *E. e. aetheria*.

All markings reduced; upf dull red pd band broken in s3; unf twin, subapical black points usually present in male, always in female; unh pd area paler, somewhat greyish. § S Balkans & NW Greece, *E. e. roosi*. Upf red band constricted in s3. § Above 1800 m in C Alps, f. *nelamus*. Smaller. Male ups gc very dark; all markings greatly reduced. Despite variation, generally easy to distinguish from similar species.

Flight-period • Early June to mid-August in one brood. Emergence date depends on region, altitude, & seasonal weather conditions.

Habitat • Sheltered gullies or hollows on open, grass-covered slopes. LHPs include mat grass (*Nardus stricta*).

Behaviour • Females appear to fly very infrequently unless disturbed, usually remaining low down in shelter of grass stems.

♀ *E. e. roosi*, NW Greece

♂ *E. e. roosi*, NW Greece

♂ *E. e. mackeri*, E France

Erebia orientalis **Bulgarian Ringlet**

Distribution • Bulgaria (Stara Planina, Rila Mts, & Pirin Mts). 1800–2600 m. Very local.

Description & variation • Resembles *E. epiphron*. Fw narrow, pointed; upf well-marked yellowy-orange/red pd spots in s4 & s5 with prominent black points, pd spots in s2 & s3 somewhat obscure, darker, smaller, lacking ocelli; uph with pd reddish-orange, black-pupilled spots; unh similar. Female markings better developed; ups black ocelli with small white pupils; unh gc yellowish grey, white-pupilled, black pd spots ringed dusky yellow.

Flight-period • Early June to early August in one brood. Emergence depends on altitude & seasonal weather conditions.

Habitat • Grassy areas near tree-line.

Erebia christi **Rätzer's Ringlet**

Distribution • SW Switzerland (Rossbodental, Eggental, Laggintal, & Zwischbergental). NW Italy (Véglia Alp). 1300–2100 m. Extremely local, generally very uncommon.

Description • Resembles *E. epiphron aetheria*, but upf blind ocelli in s3–6 clearly in a straight line.

♂ S Switzerland

Flight-period • Late June to early August in one brood.

Habitat • Steep, grassy, rocky slopes, often with scattered trees & low bushes. LHP: sheep's fescue (*Festuca ovina*).

Conservation • In decline in some localities. Protected by Swiss legislation.

♀ S Switzerland

Erebia pharte **Blind Ringlet**

Distribution • Vosges Mts, through C Alps, Julian Alps, & Tatra Mts to Carpathian Mts. 1000–2500 m, generally above 1400 m. Sporadic, locally common.

Description • Ups & uns gc brown; sharply defined, orangey-red pd markings clearly separated by brown veins, always without ocelli or black points. Male unh gc brown, female dusky yellowish brown.

♂ E Switzerland

Flight-period • Early July to late August in one brood.

Habitat • Dry or damp alpine meadows, often in woodland at lower altitudes. LHPs: grasses, including *Carex ferrunginea*, sheep's fescue (*Festuca ovina*), *F. quadriflora*, red fescue (*F. rubra*) & matgrass (*Nardus stricta*).

Erebia melampus **Lesser Mountain Ringlet**

Distribution • C Alps (Maritime Alps through Bavarian Alps & Dolomites to E Austria. 800–2400 m. Widespread, generally common.

Description • Male ups gc medium brown; upf pd orangey-red band broken by brown veins, usually with small black points in s4 & s5, sometimes also in s2 & s3; uph usually with four orange pd spots, with or without black points. (ups pattern of markings repeated on uns). Female uns gc paler, somewhat yellowish, markings similar.

Flight-period • Early July to mid-September in one brood. Emergence date depends on altitude.

Habitat • Damp or dry alpine meadows, grass-covered slopes, or grassy woodland clearings. LHPs: sweet vernal grass (*Anthoxanthum odoratum*), sheep's fescue (*Festuca ovina*), & wood meadow grass (*Poa nemoralis*).

Erebia sudetica **Sudeten Ringlet**

Distribution • France (Massif Central, Isère, & Savoie). Switzerland (Bernese Alps. 1200–2000 m. Czech Republic (Mt Praded), Slovakia, & Poland (Silesian Mts & Tatra Mts), to Romania (Carpathian Mts). 600–1200 m. Extremely local & sporadic.

Description & variation • Ups & uns gc medium brown; ups reddish, black-pupilled pd spots in a very regular series, forming almost continuous bands broken only by veins; unh series of 5–6 reddish pd spots always with black pupils (cf. *E. melampus*). § France & Switzerland. Pd markings slightly reduced.

Flight-period • Late June to August in one brood.

Habitat • Damp, flower-rich meadows, with long grasses, sheltered by woodland. LHP: sweet vernal grass (*Anthoxanthum odoratum*).

Erebia aethiops **Scotch Argus**

Distribution • Scotland. NW England (extremely local). France (Massif Central & Cévennes) through N Italy to Belgium, Latvia, S Balkans, & C N Greece. 0–2000 m, generally 300–1500 m. Widespread, locally common.

Description & variation • Upf reddish pd band usually constricted in s3. Male unh silvery-grey pd band usually well-defined, enclosing small white spots in s2–4. Female unh marginal, pd, discal bands, & basal areas clearly defined – very distinctive; ups pd ocelli variable in size & number; unh gc variable, but with all characters clearly defined. Not easily confused with similar species.

Flight-period • Late July to mid-September in one brood.

Habitat • Adapted to a wide range of habitat types. Wet or dry heathland; grassy, flower-rich meadows, often in deciduous or coniferous woodland; dry, grassy, & flowery slopes amongst scrub; on limestone or acidic soils. LHPs: a wide variety of grasses, including purple moor-grass (*Molinia caerulea*) & blue moor-grass (*Sesleria caerulea*).

♀ N Greece   ♀ N Greece

Erebia triaria de Prunner's Ringlet

Distribution • Iberia, through S France, N Italy, S Switzerland, S Austria, & Julian Alps to Albania. 400–2500 m. Sporadic, locally common.

Description & variation • Upf ocellus in s6 in a straight line with ocelli in s4 & s5; unh very dark, with obscure mottling giving a slightly roughened appearance (cf. *E. meolans*). Upf pd band variable in colour & shape; usually enclosing five ocelli, but those in s3 & s6 sometimes reduced to black points, rarely absent.

♂ E Spain

Flight-period • Mid-April to mid-July in one brood. Emergence date depends on locality & altitude.

Habitat • Bushy & grassy woodland clearings in rocky places. LHPs: grasses, including sheep's fescue (*Festuca ovina*), common meadow grass (*Poa pratensis*), alpine meadow grass (*P. alpina*), & *Stipa pennata*.

♂ E Spain

Erebia embla **Lappland Ringlet**

Distribution • Fennoscandia & NE Latvia. 100–400 m. Very & local.

Description • Male ups gc medium brown; upf orange-ringed, white-pupilled, black, oval, ocelli in s4 & s5 prominent; similar, smaller ocelli in s2 & s3 usually blind – displaced slightly towards outer margin; very small ocellus in s1b often absent; uph ocelli in s2–4 prominent; unf ocelli in s4 & s5 prominent, smaller ocellus in s3 variable; unh darker greyish-brown basal area separated from pale pd area by irregular white markings in s5–8. Female similar, ups & uns markings paler.

Flight-period • Mid-June to July in one brood.

Habitat • Grass-rich marshes bordering permanent water, usually with willow & birch scrub, often in open pine woodland. LHPs: grasses.

Erebia disa **Arctic Ringlet**

Distribution • Lappland. 300–500 m. Very local, sometimes common.

Description • Ups gc medium brown; upf orange pd spots in s2–5 fused into a prominent band; black, oval ocelli conspicuous, especially in s4 & s5.

♂ N Norway

Flight-period • Early June to late July in one brood.

Habitat • Bogs or marshes adjoining drier grassy areas or heaths with willow & birch scrub, small trees, & small areas of permanent water. LHPs: grasses.

Behaviour • For such a large & dark butterfly, shows a remarkable ability to disappear amongst tangle of grass stems when disturbed or alarmed. Only rarely seen taking nectar from flowers.

♀ N Norway

Erebia medusa **Woodland Ringlet**

Distribution • C France through S Belgium (Ardennes), N Italy to N Poland, S Balkans & N C Greece. 300–2300 m. Generally widespread, locally common.

♂ N Greece

♂ N Greece

Description & variation • Ups gc rich, medium brown; ups & uns pd yellowy-orange or orangey-red pd spots & white-pupilled, black ocelli variable, usually well-developed. Resembles *E. oeme* but uns of antennal club-tip light brown or creamy buff. § C Alps, above about 1800 m. Small; all markings greatly reduced.

Flight-period • Early May to early August in one brood according to locality & altitude.

Habitat • Grassy, flowery places, often damp, humid, woodland clearings; grassy alpine slopes amongst sparse coniferous trees. LHPs include sheep's fescue (*Festuca ovina*), red fescue (*F. rubra*), upright brome (*Bromus erectus*), & spreading millet grass (*Milium effusum*).

♀ N Greece

♀ N Greece

Erebia polaris **Arctic Woodland Ringlet**

Distribution • Mainly coastal districts of Arctic Norway & Finland. 0–400 m, generally below 200 m. Sporadic & local.

Description • Resembles *E. medusa*. Ocelli smaller; unh with pale pd band, often obscure, sometimes absent.

♂ N Norway

Flight-period • Late June to late July in one brood.

Habitat • Damp, grassy, flowery places, sometimes in open birch woodland. LHPs: grasses.

♂ N Norway

♀ N Norway

Erebia alberganus **Almond-eyed Ringlet**

Distribution • Spain (Cantabrian Mts). C Alps (Maritime Alps through Cottian Alps, S Switzerland, & Dolomites to Hohe Tauern). Italy (Ligurian Alps & Abruzzi). Republic of Macedonia (Korab Planina; 900–2200 m). Bulgaria (Stara Plannina; 1000–2000 m). Sporadic & local.

Description • Male ups & uns gc medium brown; pd orange spots enclosing black ocelli, sometimes with minute white pupils. Female markings slightly bolder; unh gc light, yellowish brown; yellowy-orange pd spots elongate – almond-shaped.

♂ SE France

Flight-period • Mid-June to late August in one brood.

Habitat • Warm, flowery meadows or grassy slopes often sheltered by woodland. LHPs include sheep's fescue (*Festuca ovina*) & sweet vernal grass (*Anthoxanthum odoratum*).

Erebia pluto **Sooty Ringlet**

Distribution • C Alps to Julian Alps. C Apennines. 1800–3000 m. Widespread.

Description & variation • A variable species, displaying a complex array of local, geographical, & intermediate forms. Male ups & uns gc very dark, sooty brown or silky black; when present, ups dark red or tawny brown pd band often obscure or diffuse, especially unh; upf & unh pd white-pupilled, black ocelli small, often absent in some populations. Female ups gc often dark, silky brown, sometimes with greyish-yellow tone; markings generally better developed; in some populations, unf obscure reddish or copper-coloured pd band shading to darker red towards wing-base; unh marginal & pd area often paler, faintly greyish. Easily separated from other species on account of size, dark colouring, & very bleak habitat in which it occurs.

Flight-period • Late June to late August in one brood. Emergence date depends on altitude.

Habitat & Behaviour • Limestone screes & moraines, largely devoid of vegetation. LHPs: grasses, including *Festuca quadriflora* & alpine fescue (*F. alpina*); these often grow as small isolated plants on screes or in small, often very isolated patches between scree slopes or in shelter of large rocks. In cool, overcast conditions, adults rest on rocks for prolonged

periods with open wings, evidently to absorb what little solar energy is available to maintain body temperature for flight. Males sometimes drink from damp soil patches. Both sexes exploit stones to shelter from persistent & usually very strong winds.

Erebia gorge **Silky Ringlet**

Distribution • Most larger massifs from N Spain (Cantabrian Mts & Pyrenees), through C Alps, C Apennines, Mte Pollino, Julian Alps, Tatra Mts, Carpathian Mts, & Dinaric Alps, to Albania (Tomor Mts) & Bulgaria (Pirin Mts & Rila Mts). 1600–3000 m. Locally common.

Description & variation • Ups gc brown; upf shiny (characteristically 'silky'), rust-coloured pd band broad, enclosing twin apical, white-pupilled, black ocelli, sometimes joined with additional ocelli in s6 forming a distinctive linear

♂ SW France

♀ SW France

group; additional smaller, often blind ocellus in s2 sometimes absent; uph orange pd band, enclosing 1–4 ocelli towards outer margin of band (regionally & locally variable); unh dark charcoal grey, marbled with lighter grey, whitish, & greyish-brown tones (regionally & locally extremely variable); paler pd band enclosing variable number of dark ocelli. Female markings better developed; unh paler, markings more distinct. Both sexes: hw outer margin with small but noticeable 'bump' at v4 – a useful distinguishing character. In some populations, ups & uns lack ocelli – prevalent form in SE France & NW Italy.

Flight-period • Late June to late August in one brood. Emergence date depends on altitude.

Habitat • Limestone screes & moraines, largely devoid of vegetation. LHPs: grasses, including alpine meadow grass (*Poa alpina*), alpine fescue (*Festuca alpina*), & *Sesleria varia*. Habitats are often shared with *E. pluto* in C Alps.

Behaviour • Resembles that of *E. pluto*.

Erebia aethiopella **False Mnestra Ringlet**

Distribution • C Alps of SE France & NW Italy.
1800–2400 m. Very local.

Description • Male ups gc brown; upf sex-brand
conspicuous; upf orangey-red pd bands broad, tapering
slightly towards inner margin, usually with small, white-
pupilled black ocelli in s4 & s5; uph pd band lacking ocelli;
unh greyish brown with paler pd band & discal area clearly
defined. Female similar; uns slightly paler, markings more
distinct.

♀ SE France

Flight-period • Mid-July to late August in
one brood.

Habitat • Sheltered sites on open, grassy,
alpine slopes. Local distribution of butterfly
& its only known LHP, *Festuca paniculata*,
appear very closely related.

Erebia rhodopensis **Nicholl's Ringlet**

Distribution • Republic of Macedonia (Stara Planina).
Bulgaria (Stara Planina, Rila Mts, Pirin Mts, & Rhodopi Mts).
NW Greece (Grammos Mt). 1800–2600 m. Very local.

Description & variation • Resembles *E. aethiopella*.
Ups & uns ocelli better developed; uph & unh pd ocelli with
small white pupils.

Flight-period • July–August in one brood.

Habitat • Open grassy slopes. LHPs: grasses

Erebia mnestra **Mnestra's Ringlet**

Distribution • C Alps. 1500–2600 m. Generally very local, often scarce.

Description • Male upf sex-brand inconspicuous; gc medium brown; reddish pd band extending to cell-end in s4 & s5; subapical ocelli generally absent or sometimes represented by two small black points; uph pd band usually restricted to s3–5 broken by veins, lacking ocelli; unf brown, marginal border well-defined, discal & basal areas uniformly reddish, unmarked; unh uniformly brown, rarely with obscure pale pd band. Female unh pale yellowish brown, with paler, greyish pd band.

Flight-period • Early July to mid-August in one brood.

Habitat • Grassy slopes. LHPs: grasses, including *Sesleria varia*.

Erebia gorgone **Gavarnie Ringlet**

Distribution • Known only from Pyrenees (Spain, Andorra, & France). 1500–2450 m. Very local.

Description • Male ups gc dark brown; pd bands dark, reddish brown, broken by dark veins, enclosing small, white-pupilled, black ocelli in s2, 4, & 5 (upf) & s2–4 (uph); upf sex-brand conspicuous; unh dark greyish brown, with paler pd band usually enclosing small, very dark ocelli. Female similar; ups & uns paler, all markings better developed; unh pale veins conspicuous.

♀ SW France

Flight-period • Mid-July to late August in one brood.

Habitat • Grassy places amongst rocks or screes. LHPs: grasses.

Erebia epistygne **Spring Ringlet**

Distribution • Spain (Montes Universales, Serrania de Cuenca, Sierra de Javalambre, Sierra de Calderesos, Sierra de Guadalajara, & Montseny Mts). 1000–1550 m. France (Cevennes to Provençe). 450–1500 m. Very local.

♀ E Spain

Description • Upf yellowish pd band tapering to inner margin, contrasting with uph orange pd band. All female markings bolder; unh gc paler, veins lined whitish.

Flight-period • Late March to late May in one brood. Emergence depends on locality, altitude, & seasonal weather conditions.

Habitat • Grassy & rocky clearings in open, usually pine woodland. LHP: sheep's fescue (*Festuca ovina*).

♀ E Spain

The following six species, commonly known as the Brassy Ringlets, comprise a closely related group. However, as far as is known, no two species of the group occupy the same habitats. In rare instances of geographical overlap in distribution, differences in altitudinal range are usually quite sufficient for reliable identification. The common generic name – brassy – derives from the very striking iridescent golden or yellowish-green sheen on the upperside, most apparent in fresh males, especially along the fore-wing costa. Members of this group are easily separated by differences in male genitalia.

Erebia tyndarus **Swiss Brassy Ringlet**

Distribution • C Alps (Pennine Alps to Brennerpass). 1200–2700 m. Widespread & very common.

Description & variation • Male ups dark brown; upf with strong, brassy, or greenish reflections, most apparent in fresh specimens, especially on costa; upf apex slightly rounded; subapical reddish patch, sometimes extending to cell in s4 & s5, usually enclosing twin, white-pupilled ocelli in s4 & s5; uph sometimes with small, reddish pd marks, lacking ocelli; unh grey gc with variable brown or greyish-brown submarginal & discal bands. Female ups & uns generally much paler; unh gc browner, all markings usually very distinct.

♂ S Switzerland

Flight-period • Generally early July to late August in one prolonged brood, sometimes extending to early October according to altitude & seasonal weather conditions.

Habitat • Open, grassy pinewood clearings at lower altitudes; grassy, stony, rocky slopes above tree-line. LHPs include matgrass (*Nardus stricta*) & sheep's fescue (*Festuca ovina*).

♂ S Switzerland

Erebia cassioides **Common Brassy Ringlet**

Distribution • Most higher mountain ranges from Spain, through S France, C Alps, & peninsular Italy, to S Balkans & NW Greece. 1600–2600 m. Very sporadic, locally often common.

Description & variation • Resembles *E. tyndarus* closely. Male ups gc slightly darker brown; upf twin, white-pupilled ocelli generally larger; uph reddish pd marks & ocelli better developed. Female resembles male but much paler; all markings more distinct. Minor regional differences in size, gc & markings largely obscured by individual & local variation.

♀ C Bulgaria

Flight-period • Late June to early September in one brood. Emergence date depends on locality & altitude.

Habitat • Grass-covered slopes or grassy places amongst rocks or screes. LHPs include sheep's fescue (*Festuca ovina*).

♀ C Bulgaria

Erebia hispania **Spanish Brassy Ringlet**

Distribution • Spain (Sierra Nevada, 1800–2900 m).
Pyrenees of Spain & France (Puerto de Portalet through
Andorra to Mt Canigou). 1650–2300 m. Locally abundant.

Description & variation • Ups gc medium brown; all
markings generally well-developed, especially upf yellowy-
orange subapical patch & enclosed twin black, white-pupilled
ocelli. Female unh, mottled yellowish grey. § Pyrenees. Slightly
smaller; all ups markings better developed.

♂ SW France

Flight-period • One brood. Sierra
Nevada: mid-June to late August according
to altitude. Pyrenees: early July to mid-
August.

Habitat • Open, grassy, & rocky slopes.
LHPs include sheep's fescue (*Festuca ovina*).

♀ E Spain

♀ E Spain

Erebia nivalis De Lesse's Brassy Ringlet

Distribution • C Switzerland (restricted to Grindelwald, 2250–2600 m). NE Italy (Atesine Alps, 2300–2450 m). Austria (Öetztal Mts to Niedere Tauern, 2100–2500 m). Locally common.

Description • Resembles *E. tyndarus*. Male upf subapical reddish patch more extensive, extending to cell in s4, s5, & v3, enclosing small, twin, white-pupilled, black ocelli; uph pd markings small or absent; unh gc lustrous, bluish grey. Female paler, browner, all markings better defined.

♀ C Switzerland

Flight-period • Early July to late August in one brood.

Habitat • Grassy places, often very small areas amongst rocky, limestone outcrops. LHP: *Festuca quadriflora*.

Erebia calcaria Lorkovic's Brassy Ringlet

Distribution • NE Italy (Mte Cavallo & Mte Santo). W Slovenia (Karwanken Alps & Julian Alps). Above 1450 m. Locally common.

Description • Resembles *E. tyndarus*. Male ups gc darker brown; upf subapical white-pupilled, black ocelli usually very small; unh gc lighter, often shiny, silvery grey.

Flight-period • Mid-July to late August in one brood.

Habitat • Grassy, rocky slopes. LHPs include sheep's fescue (*Festuca ovina*).

Erebia ottomana **Ottoman Brassy Ringlet**

Distribution • S France (Massif Central). NE Italy (Mte Baldo). S Balkans to N & C Greece. 850–2450 m, more generally 1400–2000 m. Widespread, locally abundant.

♂ N Greece

Description & variation • Ups pd markings usually well-developed. § Massif Central. Smaller. Male ups gc dark brown; upf ocelli small; unh markings usually well-developed. § Mte Baldo. Male ups gc very dark brown. Both sexes: all ups markings reduced; unh markings well-developed.

Flight-period • Mid-July to August in one brood according to altitude & season.

Habitat • Exposed, grass-covered alpine slopes or damp, grassy, woodland clearings. LHPs include sheep's fescue (*Festuca ovina*).

♀ S France

♀ (left) & ♀ N Greece

Erebia pronoe **Water Ringlet**

Distribution • Pyrenees (Spain, Andorra, & France). C Alps, through Julian Alps, Tatra Mts, Carpathian Mts, & Dinaric Alps, to S Balkans. 900–2800 m.

♂ SE France

♀ SE France

Description • Male ups gc dark brown; upf reddish pd, tapering towards inner margin, enclosing prominent, white-pupilled, black, subapical ocelli; unh gc dark, mottled grey & dark brown, silvery-grey pd band with dark violet or purple reflections, contrasting with broad, brownish discal band. Female ups gc paler, ocelli better developed; unh marginal, pd, discal bands, & basal area clearly defined.

Flight-period • Late June to late September in one brood. Emergence date depends on locality & altitude.

Habitat • Damp, grassy slopes or woodland clearings, often near small streams. LHPs include sheep's fescue (*Festuca ovina*) & *F. quadriflora*.

♀ SE France

Erebia melas **Black Ringlet**

Distribution • SW Slovenia (Mt Nanos) to S Balkans & C Greece (Mt Parnassos). 200–2800 m, usually above 1500 m in southern range. Locally common.

Description & variation • Male ups velvety black or blackish brown; upf with twin, subapical, white-pupilled, black ocelli, often with ocellus in s2; uph usually with pd ocelli in s2–4; unf as upf; unh with obscure, dark grey variegation & obscure pale marginal band. Female slightly more brown; upf & unf orangey-red pd band enclosing subapical ocelli & ocellus in s2, usually constricted in s3; all ocelli larger; unh gc brownish with greyish mottling. § Apuseni Mts (Romania). All markings better developed. Female upf bright orangey-red pd band wide, extending to s1b; ocelli larger; uph pd ocelli in s1b–s4 large, enclosed by orangey-red band; unf pd band

extending into discal area; costa, apex, & outer margin light grey; unh speckled light grey; darker discal area bordered by irregular dark line; pd ocelli large; antemarginal band paler. § S Balkans & Greece. Male uph ocelli reduced. Female gc variable, brownish to pale greyish brown; ocelli smaller; fw reddish pd bands very obscure, more usually absent.

Flight-period • Mid-July to mid-September in one brood. Emergence date depends on locality & altitude.

Habitat • Rocky, grass-covered slopes. LHPs: grasses.

♂ C Greece

♀ C Greece

♀ C Greece

Erebia lefebvrei **Lefèbvre's Ringlet**

Distribution • Spain (Picos de Europa, Sierra de la Demanda, & E Pyrenees). France (C & E Pyrenees). 1700–2700 m. Very local.

Description & variation • Resembles *E. melas* closely, but geographically widely separated. Male ups & uns gc black; upf dark, reddish pd band diffuse, variable, sometimes absent; upf & uph with 3–5 submarginal, white-pupilled, black ocelli; uns markings similar. Female gc paler; all markings better developed. Regionally & individually variable, but overall appearance always sufficient for reliable identification.

Flight-period • Late June to late August in one brood. Emergence date depends on locality & altitude.

Habitat • Steep, rocky slopes & limestone screes, interspersed with small, grassy patches or bordered by more extensive grassy & rocky slopes. LHPs: grasses.

Erebia scipio **Larche Ringlet**

Distribution • Alps of SE France & NW Italy. 1400–2500 m. Very local.

♀ S France

♀ S France

Description • Male ups gc dark brown; ups orangey-red pd bands well-developed, enclosing prominent twin, subapical, white-pupilled, black ocelli, usually with similar but much smaller spots in s3 & s4; unh dark brown with obscure paler, slightly greyish pd band. Female similar; ups gc paler brown; all markings better developed; unh pale yellowish grey with paler grey pd band. A very distinct species, unlikely to be confused with other members of genus.

Flight-period • Late July to late August in one brood.

Habitat • Steep, limestone screes, rocky slopes, or moraines with sparse grass. LHP: *Helictotrichon sedenense*, a specialized grass whose geographical & local distribution corresponds closely with that of butterfly.

Erebia stirius **Styrian Ringlet**

Distribution • NE Italy (Mte Baldo & Dolomites), through S Austria (Karawanken Alps) & Slovenia (Julian Alps) to Croatia (Velebit Mts). 700–1800 m. Very local.

Description & variation • Resembles *E. styx* closely. Ups gc brown; dull reddish pd bands prominent, crossed by brown veins; pattern of prominent, white-pupilled, black ocelli repeated on uns; unf brown marginal border slightly tapered towards s1b, without projection in s1b (distinction from *E. styx*). Male unh dark brown, with darker mottling, giving a slightly roughened appearance; pd & marginal areas paler. Female ups similar; unh gc pale mottled grey, with darker grey discal band & submarginal line. § Dolomites, high altitudes. Smaller; all markings reduced.

Flight-period • Late July to early September in one brood. Emergence date depends on altitude.

Habitat • Rocky & grassy slopes, usually on limestone. LHP: blue moor-grass (*Sesleria coerulea*).

Erebia styx **Stygian Ringlet**

Distribution • NE Italy & SE Switzerland (Mte Generoso to Dolomites). S Germany (Allgäuer Alps) through S Austria (Karwendel Mts & Zillertal Alps) to W Slovenia (Julian Alps). 600–2200 m. Very local.

♂ N Italy

Description & variation • Resembles *E. stirius* closely.
Unf brown marginal border of uniform width with short, pointed projection in s1b (distinction from *E. stirius*). Male unh less mottled, giving a smoother appearance. § S & E range. All markings generally better developed.

Flight-period • Early July to early September in one brood. Emergence date depends on altitude.

Habitat • Warm, dry, steep, limestone slopes, usually south-facing with scattered bushes & trees. LHP: *Sesleria varia*.

♀ N Italy

Erebia montana **Marbled Ringlet**

Distribution • C Alps (Maritime Alps to Dolomites); also Apuane Alps & C Apennines. 1100–2500 m. Generally widespread, sometimes common.

Description & variation • Resembles *E. styx* closely. Both sexes: unf brown marginal border internally wavy, well-defined; unh heavily mottled, pale veins conspicuous. § E Alps. Ups markings generally better developed. Male unh gc darker but variable. § Apuane Alps. Larger; unh darker.

Flight-period • Mid-July to mid-September in one brood. Emergence date depends on altitude.

Habitat • Dry or damp, grassy, flowery alpine & subalpine meadows. LHPs include matgrass (*Nardus stricta*), alpine fescue (*Festuca alpina*), & sheep's fescue (*F. ovina*).

Erebia zapateri **Zapater's Ringlet**

Distribution • Known only from E Spain (Sierra de Albarracin, Serrania de Cuenca, & Sierra de Javalambre). 1050–1650 m. Very local, often common.

Description • Resembles *E. neoridas* but not known from same region. Male ups gc dark brown; upf bright yellowy-orange pd band strongly tapered towards inner margin; subapical ocelli small but conspicuous; uph reddish pd markings small or absent; unh dark brown with obscure greyish pd band. Female similar, but brighter with all markings better developed.

♀ E Spain

Flight-period • Late July to early September in one prolonged brood.

Habitat • Open, grassy places with sparse bushes & trees. LHPs: grasses.

♂ E Spain

♂ E Spain

Erebia neoridas Autumn Ringlet

Distribution • E Cantabrian Mts, Pyrenees. Massif Central. SE France (Vaucluse, Isère, & Haute-Savoie). NW Italy (Maritime Alps to Susa Valley). Apuane Alps. C Apennines. 500–1600 m. Sporadic, locally common.

♂ SE France

Description • Resembles *E. zapateri* but not known from same region. Upf pd band dull, reddish brown (variable); uph pd orangey-red spots better developed, often forming a band with small ocelli in s2–4.

Flight-period • Early August to early October in one prolonged brood.

Habitat • Grassy, bushy places; open woodland. LHPs: grasses, including hairy finger-grass (*Digitaria sanguinalis*) & sheep's fescue (*Festuca ovina*).

♀ SE France

Erebia oeme **Bright-eyed Ringlet**

Distribution • Pyrenees through Massif Central, C Alps, Julian Alps, & Velebit Mts, to S Balkans & N Greece. 900–2600 m, generally 1500–2000 m. Very sporadic & local, especially near limits of range.

Description & variation • Male ups & uns gc dark brown; upf orangey-red subapical patch enclosing twin, white-pupilled, black ocelli; uph reddish pd spots almost circular, enclosing white-pupilled, black ocelli; pattern of markings repeated on uns with additional ocellus on unh. Female similar, all markings better developed; series of usually six ocelli on unh bold & very distinctive. Resembles *E. medusa* but uns of antennal club-tip black. Ups & uns pd markings regionally & locally variable. § Some parts of C Alps. Ups markings sometimes almost absent. § Balkans & Pyrenees. Markings often very bold. Female unh pd orange or yellowy-orange spots sometimes confluent, ocelli & ocellular white pupils large.

Flight-period • Mid-June to mid-August in one brood.

Habitat • Damp, grassy places, often in bushy, woodland clearings at lower altitudes. LHPs include alpine meadow grass (*Poa alpina*), common meadow grass (*P. pratensis*), wood meadow grass (*P. nemoralis*), red fescue (*Festuca rubra*), great pendulous sedge (*Carex flacca*), common quaking grass (*Briza media*), & purple moor-grass (*Molinia coerulea*).

♂ N Greece

♂ N Greece

♀ N Greece

♀ N Greece

Erebia meolans **Piedmont Ringlet**

Distribution · N & C Spain, through S France, C Italy, & Switzerland, to S Germany & W Austria. 600–2300 m, generally below 1500 m. Very sporadic & local, but generally common.

Description & variation · Resembles *E. triaria*, but upf small ocellus in s6 (when present) displaced towards outer margin, conspicuously out-of-line with ocelli in s4 & s5; unh very dark, smooth in appearance. § C Spain. Larger; pd bands brighter; ocelli larger. § Vosges Mts & some parts of C Alps. All markings averagely reduced but locally variable.

♀ SE France

Flight-period · Late May to mid-August in one brood. Emergence date depends on locality.

Habitat · Grassy, flowery woodland clearings. LHPs include *Agrostis capillaris*, wavy hair grass (*Deschampia flexuosa*), matgrass (*Nardus stricta*), & sheep's fescue (*Festuca ovina*).

♀ SE France

Erebia palarica Chapman's Ringlet

Distribution • NW Spain: Cantabrian Mts (provinces of Leon, Oviedo, Palencia, & Santander). 1050–1650 m. Sporadic & very local.

Description • Resembles *E. meolans*, but larger – largest European member of genus. Ups gc brown; orangey-red pd bands & white-pupilled, black ocelli very bold. Male unh dark, greyish brown, with darker, somewhat obscure discal band; 3–4 small, dark ocelli in submarginal band obscure. Female unh pattern of markings similar but paler; 4–5 submarginal ocelli better defined.

Flight-period • Late May to late July in one brood. Emergence date depends on altitude & seasonal weather conditions.

Habitat • Small, grassy, clearings amongst Broom or Mediterranean Heath. LHPs: grasses.

Behaviour • Females appear to fly very little unless disturbed.

♂ N Spain

Erebia pandrose Dewy Ringlet

Distribution • E Pyrenees. C Apennines. C Alps & Julian Alps, through Carpathian Mts & W Balkans to Bulgaria (Rila Mts, 2400–2900 m). 1600–3100 m. Fennoscandia, 0–1200 m; near sea-level on Arctic coast, progressively higher altitude towards southern limit of range. Widespread & common.

Description & variation • Very distinctive & easily identified, even in flight. All markings variable but overall character remains unmistakable. Superficially very similar to *E. sthennyo* but not known to occur in same habitat.

Flight-period • Early June to mid-August in one brood. Lappland: average emergence date mid-July.

Habitat • Open, often damp or boggy, grassy slopes & valleys. LHPs: grasses.

♂ N Norway

♂ N Norway

Erebia sthennyo **False Dewy Ringlet**

Distribution • C & E Pyrenees (Puerto de Portalet to Col du Puymorens). Above 1800 m. Very local, sometimes common.

Description • Resembles *E. pandrose* closely, but not known to occur in same habitat. Upf ocelli clearly displaced towards outer margin of pd band; unf basal & discal dark striae absent; unh markings much reduced.

Flight-period • Late June to early August in one brood.

Habitat • Grass-covered slopes with rocky outcrops.

Proterebia afra **Dalmatian Ringlet**

Distribution • Croatia (coastal districts from Zadar to ibenik). 150–500 m. NW Greece (Lake Vegoritis, Askion Mts, & Vourinos Mts). 550–1250 m. Extremely sporadic & very local, sometimes very common.

Description & variation • Very distinctive & occurs with no other species with which it is likely to be confused. All wing-characters variable.

♂ NW Greece

Flight-period • Late April to late May in one brood.

Habitat • Dry, grassy, sparsely bushy, rocky limestone slopes. Habitats are distinctive & often characterized by scattered juniper bushes as dominant shrub. LHP: sheep's fescue (*Festuca ovina*).

Behaviour • Females often visit flowers of *Globularia* & yellow daisies.

Conservation • Colonies in coastal region of Croatia vulnerable owing to close proximity to areas of intense human activity. Several have been lost in recent decades by urbanization, land management, & brush fires.

♀ NW Greece

♀ NW Greece

Maniola jurtina **Meadow Brown**

Distribution • Canary Islands. Most of Europe to S Fennoscandia, including British Isles & most Mediterranean islands (absent from E Aegean islands except Limnos & Psara). Absent from Madeira & Azores. 0–1600 m. Widespread & common.

Description & variation • Male ups medium to dark brown; upf yellowy-orange pd patch below white-pupilled, yellow-ringed, black, subapical ocellus extremely variable, usually diffuse, often absent; hw outer margin noticeably wavy; unf pd & basal areas orange, subapical ocellus repeated from upf; unh greyish brown with variable yellowish overtones; pd band paler with or without ocelli (variable in size & number when present). Female ups gc lighter; upf subapical ocellus larger, often twinned with additional, usually blind, ocellus; orange pd band well-developed, with diffuse extension to cell; uph pd area with variable orange suffusion; unf markings bolder, pd area paler yellowy-orange; unh gc yellowish brown with pale pd band, sometimes enclosing ocelli. § SW range. Larger; all markings bolder & brighter. Female unh outer discal area bordered by irregular, suffuse, yellow band. Transitional to nominate form towards eastern Mediterranean.

Flight-period • Generally late May to late October in one brood, with prolonged aestivation in southernmost range. Scotland & Scandinavia: mid-June to mid-July according to season. S Portugal: mid-April to September. Canary Islands: late March to late September according to altitude.

Habitat • Grassy & flowery places in a wide range of habitat types in dry or damp conditions. LHPs: a wide variety of grasses.

Behaviour • In hottest part of day, especially in Mediterranean region, both sexes retire to shade of bushes where they remain unless disturbed.

♀ C Greece

♀ S Spain

Maniola megala

Distribution · In Europe, known only from E Aegean island of Lesbos. Local.

Description · Resembles *M. jurtina*, but larger. Male unh pd ocelli larger, more numerous. Female ups gc darker brown; all uns markings darker, bolder, sharply defined. Easily distinguished from other species, including *M. telmessia* with which it sometimes occurs.

Flight-period · Early May to September, aestivating in summer.

Habitat · Grassy, rocky places. LHPs: grasses.

Maniola chia

Distribution · In Europe, known only from E Aegean islands of Chios & Inousses. 50–500 m. Widespread & common.

Description · Resembles forms of *M. jurtina* in eastern Mediterranean region, but not known to occur with this or related species.

Flight-period · Late May to late September in one brood, aestivating in summer.

Habitat · Grassy, rocky, & bushy places, including cultivated ground.

Maniola nurag Sardinian Meadow Brown

Distribution • Known only from Sardinia. Above 500 m. Widespread but local.

Description • Resembles forms of *M. jurtina* in W Mediterranean region; smaller, ups yellowy-orange areas more extensive, especially in female. Male upf sex-brand conspicuous. Female markings better developed, more sharply defined.

Flight-period • Late May to early August in one brood, with females appearing to aestivate in high summer.

Habitat • Grassy, flowery, bushy, & rocky places.

Maniola telmessia

Distribution • In Europe, known only from E Aegean islands of Lesbos, Samos, Ikaria, Fourni, Patmos, Leros, Kalimnos, Pserimos, Kos, Tilos, Simi, Kassos, Rhodes, Karpathos, & Kastellorizo. 0–1000 m. Locally common.

Description & variation • Resembles forms of *M. jurtina* from W Mediterranean region but not known to occur with this species in Aegean islands. § Karpathos. Male upf & female ups yellowy-orange areas more extensive.

Flight-period • Late May to late September in one brood, aestivating in summer.

Habitat • Grassy, rocky, & bushy places, including cultivated ground. LHPs: grasses.

♂ Samos, Greece

♂ Samos, Greece

Maniola halicarnassus

Distribution · In Europe, known only from Aegean island of Nissiros. 50–100 m.

Description · Resembles *M. jurtina* but male upf sex-brand large, black, & distinctly triangular. Female superficially indistinguishable from *M. telmessia*, with which it sometimes occurs.

Flight-period · Data limited: late May to early September in one brood, with prolonged aestivation.

Habitat · Bushy, grassy, flowery places.

Hyponephele lycaon
Dusky Meadow Brown

Distribution · Iberia, through S France, N Sicily, Apennines, N & C Italy, S Switzerland, & Austria, to S Finland, Balkans, & Greece. 0–2100 m. Widespread & locally common in SW & SE Europe, generally very sporadic & local elsewhere.

Description & variation · Male ups pale, greyish brown; upf sex-brand narrow, broken at v2 & v3 (cf. *H. lupina*); subapical black ocellus often blind; uph with obscure pale pd band; unh greyish brown with paler pd band. Female ups darker greyish brown; upf yellow pd band enclosing large dark ocelli in s2 & s5; suffused yellow discal patch distinctively edged by irregular brownish line; unh paler grey or greyish brown; markings better defined. Unh gc variable: in limestone habitats, characteristically whitish, brightly marbled grey.

Flight-period · June–August in one brood.

Habitat · Bushy, grassy places, usually amongst rocks. LHPs include sheep's fescue (*Festuca ovina*), red fescue (*F. rubra*), upright brome (*Bromus erectus*), & feather grass (*Stipa pennata*).

Behaviour · In hot conditions, both sexes retire to shade of bushes, rock-ledges, or tree roots. When settling, unf apical eye-spot is exposed for short time, to deflect point of attack of would-be predatory birds & lizards from more vulnerable body.

♂ NW Greece

Hyponephele lupina
Oriental Meadow Brown

Distribution · Iberia, through S France, C & S Italy, & N Sicily. W & S Balkans & Greece, including Lesbos, Chios, Kos, Rhodes, & Crete. 0–2000 m. Very local, generally uncommon.

Description & variation · Resembles *H. lycaon*, but larger. Male ups dark brown with golden reflections; upf subapical ocellus obscure; upf sex-brand broad, not broken by veins (cf. *H. lycaon*). Female upf discal area without yellow or orange suffusion; yellowish pd band usually reduced to patches enclosing large, prominent, dark ocelli. Both sexes: hw outer margin wavy. Regionally & locally variable in size, male ups gc, length, & width of sex-brand.

Flight-period · Mid-May to mid-August in one brood, with females aestivating in high summer or prolonged & abnormally hot conditions.

Habitat · Hot, dry, grassy, & bushy places. LHPs: grasses.

Behaviour · Both sexes retire to shade of bushes in hot conditions

Aphantopus hyperantus **Ringlet**

Distribution · N Spain (Cantabrian Mts to E Pyrenees), through much of Europe including British Isles, to C Fennoscandia, Balkans, & N Greece. 0–1600 m. Locally common.

Description & variation · Ups gc dark greyish brown; yellow-ringed, white-pupilled, black pd ocelli variable in number & size; unh pd ocelli distinctive & characteristic; unh without pd & antemarginal metallic lines; yellow-ringed ocelli in s5 & s6 displaced basad; s4 without ocellus (cf. *C. oedippus*). § N Britain. Smaller; ups & uns gc tending to greyish yellowy-brown; ocelli often small; unh ocelli sometimes replaced by small white dots. § SE Europe. Large; all markings bold; uns gc brighter, with brassy golden reflections.

Flight-period · Mid-June to late August in one brood.

Habitat · Damp or dry grassy, bushy places or woodland clearings. LHPs: a wide variety of grasses.

Behaviour · Both sexes, especially females, confine much of their activity to shaded, bushy margins of woodland or large thickets; these often host preferred, shaded specimens of LHPs amongst which females usually deposit their eggs or where emerging females attract males. Both sexes strongly attracted to nectar-rich plants, especially bramble (*Rubus fruticosus*).

♀ N Greece

♀ N Greece

Pyronia tithonus **Gatekeeper**

Distribution • Iberia, through most of Europe, including Corsica, Sardinia, S Ireland, S England, N Germany, & C Poland, to Balkans & Greece. 0–1700 m. Generally widespread & often abundant, more sporadic in some parts of central range.

Description & variation • Male ups gc brown; orange bands prominent, extending to wing-base on fw; upf sex-brand conspicuous, distinctive (not segmented – cf. *P. cecilia*); subapical twinned, white-pupilled, black ocelli conspicuous; unh yellowish grey, with distinctive pale costal mark & pd band. Female similar, but upf orange area not disrupted by

♂ E France

sex-brand; all other markings bolder. § Hotter localities of S Europe. Unh gc paler, often yellowish buff.

Flight-period • Early July to early September in one brood.

Habitat • Grassy, flowery, bushy places, often damp & humid, usually associated with deciduous or pine woodland. LHPs: wide variety of grasses.

Pyronia cecilia **Southern Gatekeeper**

Distribution • Iberia, including Balearic Islands, through S France, Corsica, Italy, Sardinia, Elba, Giglio, & Sicily, to W & S Balkans & Greece, including Corfu & Levkas. 0–1200 m. Very sporadic & very local throughout range.

Description & variation • Resembles *P. tithonus*. Ups orange markings more extensive, extending to cell on upf & uph; unh markings more grey, paler. Male upf androconial patch segmented (cf. *P. tithonus*). § Western range. Unh marbled pale grey & white. § Eastern range. Unh duller grey with brownish tones, often with diffuse, yellow pd shading.

♀ E Spain

Flight-period • Early June to mid-August in one brood.

Habitat • Hot, dry, sparsely grassy, rocky scrubland. LHPs: grasses.

Behaviour • In hot conditions, both sexes seek shade of bushes or rocks.

Pyronia bathseba Spanish Gatekeeper

Distribution • Iberia (south of Cantabrian Mts), through S France (E Pyrenees to Maritime Alps). 300–1700 m. Widespread, locally common.

♂ E Spain

♂ E Spain

Description • Superficially similar to *P. tithonus* & *P. cecilia*. Ups orange pd bands well-developed, especially in female; uph pd ocelli prominent. Male upf discal area dark, obscuring sex-brand; unh dull, greyish brown with very distinctive, pale creamy-yellow, well-defined discal band; 5–6 orange, yellow-ringed, white pupilled, black pd ocelli well-defined.

Flight-period • Late April to July in one brood according to locality.

Habitat • Grassy, bushy places, often in light woodland. LHPs: grasses, including slender brome (*Brachypodium sylvaticum*).

♀ S Spain

Coenonympha tullia **Large Heath**

Distribution · Ireland, Wales, NE England, Scotland, & E France, to Fennoscandia & C & NW Balkans. 0–1200 m. Colonies generally extremely small & widely dispersed in western & southern range.

Description & variation · Ups gc dull, yellowish ochre; fringes whitish; wing-margins lightly suffused greyish white; ups pd ocelli variable in size & number, but subapical spot always present; upf often with pale pd bar near apex. Male uph with diffuse, creamy mark near costa, extending across disc in female; unh basal area greyish or greyish brown, separated from more brown pd area by irregular creamy or whitish marks; white-ringed, black ocelli in pd band variable in size & number. Female similar; ups brighter, all markings better developed, uph basal area usually greyish brown. All wing-characters variable, sometimes quite markedly so, even within a small part of geographical range. However, easily separated from all other species likely to be found within same, usually very distinctive, habitat.

Flight-period · Mid-June to early August in one brood. Emergence date depends on locality.

Habitat · Peat bogs, mosses, & wet heaths, often with small, scattered birch & pine trees & patches of shrubs such as willow & bilberry. Most habitats are easily recognized by their distinctive moorland grasses & sedges, several of which

comprise LHPs, especially common cotton grass (*Eriophorum angustifolia*), *E. vaginatum*, *Carex rostrata*, & white-beaked sedge (*Rhynochospora alba*).

Conservation · As with all wetland species, very sensitive to interference of habitat, especially land drainage, possibly even in areas well-removed from habitat.

Coenonympha rhodopensis
Eastern Large Heath

Distribution · Italy (C Apennines & Mte Baldo). Romania (Retezat Mts). W Croatia to Bulgaria & N Greece. 1400–2200 m. Local, often common.

Description & variation · Superficially very similar to *C. tullia*. Ups brighter orangey yellow, lacking marginal suffusion. Male unf pale pd stripe near apex usually absent. § Italy & some localities of Bulgaria & Greece. Unh ocelli of equal size, in a complete series.

Flight-period · Mid-June to late July in one brood.

Habitat · Open, usually somewhat dry grassland above tree-line, sometimes damp forest clearings. LHPs: grasses.

♂ N Greece

Coenonympha pamphilus **Small Heath**

Distribution • Most of Europe including British Isles & most Mediterranean islands. Absent from Canary Islands, Azores, Madeira, Orkney Islands, Shetland Islands, Crete, & SE Aegean islands. 0–1950 m. Widespread, very common.

Description & variation • Ups gc yellowish buff to dull yellowy orange; marginal borders brownish, variable in colour & width; unh gc variable, grey to brownish; unh pd ocelli sometimes vestigial or absent. § Mediterranean region. Ups dark submarginal borders wider, especially in summer broods; unf with diffuse, black submarginal stripe in s1–5 & oblique reddish-brown pd bar near apex; unh basal & discal areas light sandy brown, sometimes with well-defined, darker submarginal border; pd area pale creamy buff; ocelli vestigial, often absent.

Flight-period • Northern range: May–October in 2–3 broods. Southern range, low altitudes: February–November in 3+ broods.

Habitat • Grassy places in a wide variety of habitat types. LHPs: a wide variety of grasses, including sheep's fescue (*Festuca ovina*), red fescue grass (*F. rubra*), annual meadow grass (*Poa annua*), sweet vernal grass (*Anthoxanthum odoratum*), crested dog's tail grass (*Cynosurus cristatus*), cock's-foot grass (*Dactylis glomerata*), & matgrass (*Nardus stricta*).

♂ S Spain

Coenonympha thyrsis **Cretan Small Heath**

Distribution • Known only from Crete. 0–1800 m.
Widespread, locally common.

Description & variation • Resembles *C. pamphilus* but
does not occur with this species. Ups gc yellowish, well-defined
discal areas slightly darker; internal edge of brown marginal
borders distinctly scalloped; uph small, dark pd ocelli variable,
usually reduced to one dark point in male anal angle; upf dark
brown subapical ocellus prominent. § Higher altitudes.
Smaller; ups black marginal borders narrower; uns markings
less prominent.

Flight-period • Generally early May to
early July in 1+ broods, according to altitude.
Records extend to early October.

Habitat • Grassy, often dampish areas in a
wide variety of habitat types. LHPs: grasses.

♂ Crete, Greece

Coenonympha corinna **Corsican Heath**

Distribution • Known only from Corsica, Capraia, &
Sardinia. 0–2000 m, generally below 1200 m. Widespread,
locally common.

Description • Resembles *C. elbana* but not known to occur
with this species. Ups gc orangey yellow; dark brown borders,
uph brown costal area & brown veins in pd area distinctive;
upf subapical orange-ringed ocellus, sometimes with very

small white pupil. Female uph basal area
with darker shading; small brown pd points
usually present in s2–4.

Flight-period • Mid-May to August in
2+ broods.

Habitat • Grassy, bushy, & rocky places in
open woodland or margins of cultivation.
LHPs: grasses.

♂ Corsica, France

Coenonympha elbana **Elban Heath**

Distribution • W Italy (coastal area of Tuscany & islands of Elba, Giglio, & Giannutri). 0–800 m. Locally common.

Description • Resembles *C. corinna* closely but not known to occur with this species. Unf dark antemarginal & submarginal lines prominent; unh pale pd band & external dark border almost linear in s2–4; ocelli prominent, usually in complete series.

Flight-period • Early May to September in 3+ broods.

Habitat • Grassy, bushy places. LHPs: grasses.

Coenonympha dorus **Dusky Heath**

Distribution • Iberia, through S France to NW Italy (Maritime Alps) & Apennines. 100–1700 m. Widespread & common in Iberia, sporadic & very local in Italy.

Description & variation • Male upf smoky brown; yellow-ringed ocellus (often twinned) conspicuous; uph gc dull yellowy orange; discal line paler; costa smoky brown; basal area shaded orangey brown; marginal & antemarginal brown lines distinctive; pd ocelli in s1c–4 in a proximally convex curved line – characteristic; ocellus in s6 obscure but distinctive; all uns markings very distinctive & characteristic. Ups coloration & uns markings locally & regionally variable. § N Iberia & locally in E Spain & S France. Male ups often darker, smoky brown, ocelli smaller, often reduced in number; uns markings less contrastive, ocelli smaller.

Flight-period • Early June to mid-August in one brood.

Habitat • Dry, grassy, & rocky places, usually amongst scrub, often in dry, open pine woodland. LHPs include brown bent-grass (*Agrostis canina*), white bent-grass (*A. alba*), & sheep's fescue (*Festuca ovina*).

♂ E Spain

Coenonympha arcania **Pearly Heath**

Distribution • N & E Spain, through France & Italy to
S Norway (very local in Oslo Fjord), S Sweden (very local),
Balkans, & C Greece. 50–1800 m. Generally widespread &
common.

Description & variation • Upf disc orange, contrasting
with well-defined broad, brown, marginal border; uph smoky
brown with narrow, orange band in anal angle; unh irregular,
creamy-white pd band & series of 5–6 ocelli prominent
(cf. *C. darwiniana*).

Flight-period • Mid-May to mid-August
in one prolonged brood.

Habitat • Grassy, flowery, bushy places,
usually associated with damp or dry
woodland clearings or margins. LHPs:
meadow grass (*Poa pratensis*), melick grass
(*Melica ciliata*), & Yorkshire fog (*Holcus
lanatus*).

♂ NW Greece

Coenonympha darwiniana **Darwin's Heath**

Distribution • C Alps, SE France, S Switzerland (southern
alpine slopes), & N Italy (Venosta to Dolomites). 800–2100 m.
Very sporadic & local.

Description • Resembles *C. arcania*, but
generally smaller. Unh white or creamy-
yellow pd band much narrower, ocelli ringed
yellow (cf. *C. gardetta*).

Flight-period • Early June to August in
one brood.

Habitat • Flowery, grassy places. LHPs:
grasses.

♂ S Switzerland

Coenonympha gardetta **Alpine Heath**

Distribution • France (Massif Central; very local). C Alps through Julian Alps & Dinaric Alps to Albania. 800–2900 m, generally above 1500 m. Generally sporadic & local.

Description • Ups greyish, sometimes with faint orange basal & discal flush upf; unh superficially similar to that of *C. darwiniana* but ocelli smaller, not ringed yellow, usually fully enclosed in broader, white pd band; narrow antemarginal band pale yellow.

Flight-period • Late June to mid-September in one brood. Emergence date depends on altitude.

Habitat • Open alpine meadows at higher altitudes, grassy slopes with sparse bushes & trees at lower altitudes. LHPs: grasses.

Coenonympha leander **Russian Heath**

Distribution • Locally common. Romania (S Carpathian Mts). Republic of Macedonia. Bulgaria. N Greece. 350–1900 m. Very sporadic & local.

Description & variation • Male ups gc dark brown; upf disc orange, shading into pd area; uph gc brown; anal angle orange; orange-ringed, brown ocelli in s2 & s3; unh disc light orangey brown with greyish shading; margin very pale; submarginal band orange, separated from creamy-white pd band by six cream-ringed, white-pupilled, black ocelli. Female closely similar; markings better developed. § Pindos Mts (Grammos massif to Tzoumérka Mts; 1000–1900 m), *C. l. orientalis*. Resembles nominate form but unh ocelli internally bordered by prominent white pd band.

Flight-period • Generally late May to mid-July in one brood. Records span mid-April to early August, according to locality & altitude.

Habitat • Warm, grassy, flowery, bushy, woodland margins or forest clearings. LHPs: grasses.

Behaviour • Flight mostly confined to warm, bushy, sheltered places. Both sexes strongly attracted to nectar of raspberry & bramble. Female egg-laying sites appear confined to grasses in well-shaded situations, often inside tangled thickets of bramble & other shrubs.

♂ *C. l. leander*, N Greece

♂ *C. l. orientalis*, N Greece

Coenonympha glycerion **Chestnut Heath**

Distribution • N & E Spain, through E Pyrenees, Massif Central, Italy (Ligurian Alps, Cottian Alps, & C Apennines), E France (Provençe to Ardennes), & Switzerland, to S Finland, Balkans, & N Greece (Rhodopi Mts). 250–2100 m, generally below 1800 m. Locally common.

Description & variation • Ups tawny brown, often with orange basal flush through cell & costa; uph with narrow, well-defined orange band in anal angle; unf orange with grey marginal border; unh white discal marks in s1c & s4 distinctive & characteristic; pd white-pupilled, dark ocelli variable in size & number; orange antemarginal line in anal angle variable.

♀ N Greece

Female upf gc orange; fine marginal border dark, separated from similar submarginal band by orange, antemarginal band; uph gc brown with greyish overtone; orange antemarginal border; sometimes with 2–3 orange-ringed, dark ocelli near anal angle; uns similar to that of male. § Spain (S Cantabrian Mts & Pyrenees to Montes Universales, 600–1600 m), *C. g. iphioides*. Slightly larger; all wing-characters much better developed; unf with 1–2 small subapical ocelli. Intermediate forms occur at higher altitudes in Pyrenees & Montes Universales. Darker, greyer forms occur on bogs & damp heaths.

Flight-period • Late May to August in one brood. Emergence date depends on locality.

Habitat • Dry or damp, grassy, bushy places often in woodland clearings or margins. LHPs: wide variety of grasses.

♂ *C. g. iphioides*, E Spain

Coenonympha hero **Scarce Heath**

Distribution • N & E France through N & C Switzerland to S Fennoscandia in very few, widely dispersed, & mostly extremely small colonies. 50–700 m. Very uncommon in some colonies.

Description • Ups dark brown; uph with 2–3 ocelli in anal angle of male, 3–4 in female, with additional ocellus in upf subapex.

♀ SE France

Flight-period • May to early July in one brood.

Habitat • Damp or wet grassy meadows. LHPs: grasses.

Behaviour • Rests & roosts amongst grass stems with closed wings.

Conservation • Very rare & endangered – probably most threatened of all European butterflies. In rapid decline in most of C Europe. Exploitation of habitats for commercial purposes, e.g. afforestation, continues despite legislative protection of the butterfly.

Coenonympha oedippus **False Ringlet**

Distribution • W France to C Balkans in very few, very small, & widely dispersed colonies. 150–500 m.

Description • Ups dark brown, with prominent yellow-ringed, white-pupilled, dark ocelli in pd area of hw of female; unh gc yellowish brown; uns (fw & hw) pale, silvery-grey antemarginal line distinctive; unh ocelli prominent, almost linear in s1b–5, internally bordered by irregular pale pd band; ocellus in s5 smaller (cf. *A. hyperanthus*); ocellus in s6 displaced towards wing-base (cf. *C. hero*). Female markings bolder, with additional pd ocelli unf. Overall character distinctive. Similar to, but not easily confused, with *A. hyperanthus*.

Flight-period • Early June to early August in one brood.

Habitat • Wet, grassy places, invariably associated with rivers or lakes. LHPs: common meadow grass (*Poa pratensis*), annual meadow grass (*P. annua*), & *P. palustris*.

Conservation • A seriously threatened species: circumstances pertaining to *C. hero* apply equally to *C. oedippus*.

Pararge aegeria **Speckled Wood**

Distribution • Madeira (probably introduced about 1976). Most of Europe, including British Isles, Channel Islands, Balearic Islands, Corsica, Sardinia, Sicily, Corfu, Crete, Thassos, Lesbos, Samos, Kithira, Kos, & Karpathos. Absent from Canary Islands & Azores. 0–1750 m. Widespread & common.

♂ N Switzerland

♂ SW France

Description & variation • § *Pararge a.* f. *aegeria*. Ups gc brown, with distinctive pattern of orange spots; upf subapex with small, white-pupilled, black ocellus; uph orange spots in s2–4 enclosing similar but larger ocelli. Female markings bolder. Both sexes: fw outer margin slightly concave below subapex (cf. *P. xiphia*). § Britain, NW & SE France to Fennoscandia, Balkans, & Greece, *Pararge a.* f. *tircis*. Ups & uns gc slightly darker, greyer tone; orange markings replaced by creamy white or yellow. Intermediate forms are common in warmer regions.

Flight-period • Northern range: late March to June & late June to early October in two broods. Warmer regions: late February to early October in 2–3 broods. Emergence date depends on locality & altitude.

Habitat & behaviour • Wide variety of woodland types, often sparsely wooded margins of rivers or small streams. Shows marked preference for shady areas with dappled sunlight, where males often sit on sunlit leaves with half-open wings. Males very territorial, defending domain vigorously from intruders of same species & sex. LHPs: wide variety of woodland grasses.

Slow but progressive extension of range in NE Scotland in recent decades into previously unoccupied but suitable-looking habitat suggests changing climatic influence.

♀ SW France

Pararge xiphioides
Canary Islands' Speckled Wood

Distribution • Canary Islands (Gomera, La Palma, Tenerife, & Gran Canaria). 200–2000 m. Local, not common.

♂ Tenerife, Canary Isles

Description • Resembles *P. aegeria*. Fw outer margin linear; unh with white band extending from costa to cell. Very distinctive – no related species occurs in Canary Islands.

Flight-period • Throughout year in several broods.

Habitat • Laurel & chestnut forests, less often pinewoods. LHPs: grasses.

Pararge xiphia Madeira Speckled Wood

Distribution • Madeira. 0–1000 m. Local, scarce.

Description • Fw outer margin slightly convex below subapex; unh with distinctive small white mark on costa (cf. *P. aegeria*).

Flight-period • Throughout year in several broods, with no apparent discontinuity but with reduced abundance in June–August.

Habitat • Laurel & chestnut forests. LHPs: grasses, including slender false brome grass (*Brachypodium sylvaticum*), Yorkshire fog (*Holcus lanatus*), & giant bent-grass (*Agrostis gigantea*).

Lasiommata megera **Wall Brown**

Distribution · Most of Europe, including Ireland, Wales, England, SW Scotland (extremely rare & local), Baltic islands, & most Mediterranean islands. 0–2300 m. Generally widespread & common.

Description & variation · Upf with two conspicuous, transverse cellular bars (cf. *L. maera*). Unh brownish in N Europe, greyish in S Europe. § Tyrrhenian islands (including Corsica & Sardinia), f. *paramegera*. Resembles *L. m. megera*, but smaller; ups markings reduced, especially upf dark pd lines in s1b & s2; uph irregular pd band absent (cf. *L. m. megera*). Intermediate forms occur in Balearic Islands & Sicily.

Flight-period · Early April to October in 2–3 broods, according to locality & altitude.

Habitat · Grassy, rocky places, flowery meadows, & woodland clearings. LHPs: wide variety of grasses.

Behaviour · Males are very alert & easily disturbed from paths, stones, rock faces, or walls upon which they often bask. Shelter provided by overhanging rocks or shaded tree-roots are favoured roosting sites in hot climates. Males often assemble on mountain summits.

♀ NW Greece

♂ N Greece

♂ NW Greece

Lasiommata maera **Large Wall Brown**

Distribution · Most of Europe, including Sicily & several
Ionian & Aegean islands (except Crete). 0–2000 m. Widespread
& common.

Description & variation · Upf with one transverse
cellular bar (cf. *L. megera*); uph without transverse discal line
(cf. *L. petropolitana*); unh gc variable, light grey to greyish
brown. § Fennoscandia. Smaller; ups dark suffusion more
extensive, but upf subapical yellowy–orange patch retained.
§ Iberia & Mediterranean region. Female upf orange markings
more extensive.

♀ NW Greece

Flight-period · N Europe: mid-June to
late August in one brood. S Europe: late
April to June & June to late September in
two broods.

Habitat · Dry, often hot, grassy,
rocky/stony places, including steep slopes &
screes. LHPs: wide variety of grasses.

Behaviour · Males sometimes assemble
on mountain summits – recorded at 2900 m
in S Spain (Sierra Nevada). Often basks on
rocks in early morning & late afternoon sun.
Very wary & easily disturbed. Often retires to
shelter provided by overhanging rock-ledges.
Roosts in rock crevices.

♀ NW Greece

Lasiommata petropolitana
Northern Wall Brown

Distribution • Most higher mountain ranges from Pyrenees, through C Alps to Carpathian Mts & N C Greece. 500–2250 m. Fennoscandia. 100–1200 m. Sporadic & generally very local.

♂ N Greece

Description • Ups, dull, greyish brown; uph with irregular, transverse discal line (cf. *L. maera*).

Flight-period • Late April to early August in one brood.

Habitat • Grassy & stony places, sandy, rocky banks, or gullies in woodland clearings or margins. LHPs: grasses, including small wood-reed (*Calamagrostis epigejos*), sheep's fescue (*Festuca ovina*), & cock's-foot (*Dactylis glomerata*).

Behaviour • Adults often rest on bare ground or fallen tree-trunks. Usually shelters or roosts under rock-ledges or exposed tree-roots.

♀ N Greece

♀ N Greece

Lopinga achine **Woodland Brown**

Distribution • N Spain, through C France, NW Germany, & N Italy, to SE Sweden, S Finland, & C Balkans. 200–1500 m. Generally very sporadic & local.

♂ N Switzerland

Description • Ups & uns gc greyish brown, with large, distinctive, yellow-ringed, brown ocelli; uns markings similar, very bold, with pd whitish bands. Unlike any other European satyrid.

Flight-period • Early June to late July in one brood.

Habitat • Dry or damp, grassy, bushy, woodland clearings. LHPs include slender brome (*Brachypodium sylvaticum*) & tor grass (*B. pinnatum*).

Conservation • Reputedly in decline in France, Switzerland, & N Italy.

♂ S France

Ypthima asterope **African Ringlet**

Distribution • Greece: Aegean islands of Samos, Rhodes, Simi, & Kastellorizo. 0–250 m. Local.

Description • Ups greyish brown; upf with prominent, twin white-pupilled, yellow-ringed, subapical ocellus, repeated on unf; uph with white-pupilled, yellow-ringed ocellus in anal angle; unh mottled greyish white with submarginal, yellow-ringed ocelli.

Flight-period • Early April to late October in 2+ broods.

Habitat • Hot, stony, grassy places, often dry river beds. LHPs: grasses.

Kirinia roxelana **Lattice Brown**

Distribution • SW Croatia (Dalmatian coast), through S Romania to European Turkey & Greece, including Corfu, Levkas, & most E Aegean islands except Crete. 0–1750 m. Widespread but local, often common.

Description • Ups gc greyish brown. Male upf with dark, distinctive sex-brand – visible in flight. Female unh gc paler, markings bolder.

Flight-period • Late April to September in one brood. Appears to aestivate in hottest/driest summer months.

Habitat • Generally hot, dry, bushy places, often in open pinewoods; less often in cooler, damper conditions at higher altitudes. LHPs: grasses.

Behaviour • Both sexes appear to spend much time in interior, often deep-shade, of bushes, thickets, or small trees, quickly retreating to such cover when disturbed. Males sometimes roost amongst rocks in exposed places, females more usually in bushes. In hot, calm conditions, both sexes sometimes gather at dusk in large numbers on stony beds of dry water-courses, from which they are easily disturbed during early hours of darkness. Feeds & rests with wings closed.

♂ N Greece

♀ NE Greece

Kirinia climene **Lesser Lattice Brown**

Distribution • S Balkans & NW Greece. 700–1600 m. Very sporadic & extremely local.

Description & variation • Uph orange pd band variable, sometimes absent. Female unh gc usually yellowish, less often greyish.

♂ NW Greece

Flight-period • Mid-June to late July in one brood.

Habitat • Grassy, bushy clearings (often with raspberry (*Rubus idaeus*)) in damp or dry mature, deciduous, or mixed woodland. LHPs: grasses, including slender false brome (*Brachypodium sylvaticum*).

Behaviour • Adults often visit forest canopy, where they roost. Feeds & rests with wings closed.

♀ NW Greece

Hesperiidae

This family of small butterflies – the Skippers – is readily distinguished from all others by a very robust appearance arising from a large head and thorax, large eyes, and widely separated antennae. Rapid flight, often close to the ground, is characteristic of the family. A few species are extremely wary and difficult to approach, but when disturbed will often return to their original resting site after a brief period. In hot conditions, males of some species may gather in large numbers to drink from damp ground.

Pyrgus malvae Grizzled Skipper

Distribution • Most of Europe, including S Britain. Absent from Mediterranean islands except Sicily & Lesbos. 0–1900 m. Widespread & common.

Description • Ups very dark gc contrasting sharply with well-defined white markings.

Flight-period • April to early July in one brood, or April to early June & late July to August in two broods. Emergence date depends on locality & altitude.

Habitat • Grassy, flowery places. LHPs: principally cinquefoils, including tormentil (*Potentilla erecta*); also, wild strawberry (*Fragaria vesca*), agrimony (*Agrimonia eupatoria*), & blackberry (*Rubus fruticosus*).

♂ N Greece

♂ N Greece

♀ NW Greece

Pyrgus alveus **Large Grizzled Skipper**

Distribution • Most of Europe. 800–2000 m. Very sporadic & local.

Description & variation • Ups gc dark, greyish brown, sometimes with whitish or golden scaling in fresh female specimens. § Norway & Sweden (100–1100 m), *P. a. scandinavicus.* Resembles nominate form but smaller; ups white markings more clearly defined.

♀ N Spain

Flight-period • June to mid-August in one brood. Emergence date depends on locality & altitude.

Habitat • Dry or damp grassy, flowery places. LHPs include barren strawberry, (*Potentilla sterilis*), wild strawberry (*Fragaria vesca*), agrimony (*Agrimonia eupatoria*), & rockrose (*Helianthemum nummularium*).

♀ N Greece

Pyrgus armoricanus
Oberthur's Grizzled Skipper

Distribution • S Europe, including Sardinia, Corsica, Sicily, Kithira, & Crete, to Denmark & S Sweden. 50–1700 m. Very sporadic & local in northern range.

Description • Resembles *P. alveus*, but ups light coloration & yellowish unh gc characteristic & distinctive.

Flight-period • Generally late April to June & July–September in two broods (in Sardinia, most abundant in October). Northern range: late June to July in one brood.

♂ N Greece

Habitat • Grassy, rocky gullies/slopes, often hot, dry, bushy places with an abundance of flowers. LHPs include spring cinquefoil (*Potentilla tabernaemontani*), creeping cinquefoil (*P. reptans*), wild strawberry (*Fragaria vesca*), & rockrose (*Helianthemum nummularium*).

Behaviour • Strongly attracted to flowers of thyme (*Thymus*) & a yellow species of *Achillea*.

Pyrgus foulquieri
Foulquier's Grizzled Skipper

Distribution • Spain (Catalonia; very local). S France to NW Italy (Maritime Alps) & Apennines (Mte Sibillini & Mte Aurunci). 500–1800 m. Very local.

Description & variation • Resembles *P. alveus*. Uph pale markings better developed; unh gc yellowish brown. § C Italy. Smaller; ups paler.

Flight-period • Mid-July to August in one brood.

Habitat • Grassy, flowery places. LHPs: cinquefoils (*Potentilla*).

Pyrgus warrenensis **Warren's Skipper**

Distribution • Alps of E Switzerland (Albulapass, Julier Pass, & Bernina Pass), NE Italy (Ortler Alps & Dolomites), & W Austria (Brenner Pass & Hohe Tauern). 1800–2600 m. Widespread but very local.

Description • Resembles *P. alveus* but smaller; markings less prominent. *P. alveus* does not occur in same habitat.

Flight-period • July–August in one brood.

Habitat • Sheltered hollows, gullies, & slopes with short turf & thyme (*Thymus*).

Pyrgus serratulae **Olive Skipper**

Distribution • Spain to C E & SE Europe. Not reported from Mediterranean islands. 50–2400 m. Sporadic in western range; generally widespread & locally common.

Description • Uns gc & markings distinctive & characteristic. Female ups often with golden or whitish scaling.

Flight-period • Late April to July in one brood.

♀ C Greece

Habitat • Grassy, flowery places; damp, woodland clearings; hot, dry scrubland at low altitudes. LHPs: cinquefoils, including sulphur cinquefoil (*Potentilla recta*), spring cinquefoil (*P. tabernaemontani*), & creeping cinquefoil (*P. reptans*).

♀ C Greece   ♂ N Greece

Pyrgus carlinae Carline Skipper

Distribution • Portugal, through C France, NW Italy, & S Germany, to W Hungary. Generally 800–1900 m; in eastern range (sporadic & local), 300–1300 m; Iberia (widespread & common), 800–1600 m.

♂ E Spain

Description & variation • § Western range, *P. c. cirsii*. Ups white markings better developed; unh gc striking yellowy brown to rich reddish brown. Male genitalia distinctive; distributional relationship with nominate form uncertain.

Flight-period • Late July to August in one brood.

Habitat • Flowery, grassy places. LHPs: cinquefoils, including creeping cinquefoil (*P. reptans*), hairy cinquefoil (*P. hirta*), spring cinquefoil (*P. tabernaemontani*), common tormentil (*P. erecta*), & barren strawberry (*P. sterilis*).

♂ E Spain

Pyrgus onopordi **Rosy Grizzled Skipper**

Distribution • Portugal & Spain, through S France to Italy.
0–2000 m. Generally widespread, but sporadic in NE Italy.

Description • Unh large, white, anvil-shaped spots in s4
& s5 distinctive & characteristic.

Flight-period • April to early October in 2–3 broods.

Habitat • Flowery meadows, sheltered streams, & rocky
gullies in open grassland. LHPs include dwarf mallow
(*Malva neglecta*).

♂ E Spain

♀ E Spain

Pyrgus cinarae **Sandy Grizzled Skipper**

Distribution • Spain (Montes Universales). 900–1200 m.
Extremely local & apparently very rare. Albania. Republic
of Macedonia. Bulgaria. N & C Greece. European Turkey.
750–1600 m. Very sporadic & local.

Description • Spain, *P. c. clorinda*: ups & uns gc more
yellowish, especially male unh.

Flight-period • SE Europe: mid-June to early August in one
brood. Spain: mid-July to early September.

Habitat • Open, dry, grassy, flowery places amongst scrub &
rocks, sometimes in light woodland. LHPs: cinquefoils,
including sulphur cinquefoil (*Potentilla recta*).

Behaviour • In SE Europe, males often visit damp places
to drink.

♂ N Greece

♀ N Greece

Pyrgus sidae Yellow-banded Skipper

Distribution • W Spain (Sierra de Gredos). 700–1300 m. SE France, NW Italy. 100–1400 m. Slovenia to N Greece & European Turkey. 50–1750 m, generally above 600 m. Sporadic, locally common.

Description • Spain to Italy, *P. s. occiduus*: smaller; unh markings generally less distinct, yellow discal band generally paler.

♂ NW Greece

Flight-period • One brood. Generally mid-May to late June. Near sea-level in NE Greece: early April to May.

Habitat • Grassy places containing an abundance of flowers, which often include vetches (especially *Vicia cracca* & yellow-flowered *Achillea* species – much favoured nectar sources). LHPs include sulphur cinquefoil (*Potentilla recta*).

Behaviour • Males often visit damp ground to drink.

♂ NW Greece

Pyrgus carthami **Safflower Skipper**

Distribution • N Portugal & Spain, through peninsular Italy to S Lithuania, European Turkey, & N Greece. 600–1800 m. Sporadic & very local.

Description • Uph white, elongate submarginal markings usually prominent; unh white markings narrowly bordered pale grey – distinctive & characteristic.

♂ N Greece

Flight-period • Late June to September in one brood.

Habitat • Sheltered grassy, flowery places, often amongst rocks, bushes, or in open woodland. LHPs include grey cinquefoil (*Potentilla cinerea*), *P. arenaria*, hairy cinquefoil (*P. hirta*), & spring cinquefoil (*P. tabernaemontani*).

♀ N Greece

♀ N Greece

Pyrgus andromedae
Alpine Grizzled Skipper

Distribution • C European Alps to Julian Alps & Carpathian Mts. S Bosnia-Herzegovina. SW Serbia. Republic of Macedonia. 1500–3000 m. C & N Norway & N Sweden. 25–1000 m. Generally sporadic & local, especially in S Balkans.

♂ N Norway

♀ N Norway

Description • Unh prominent white spot & white streak in s1c gives a striking impression of an exclamation mark – distinctive & characteristic.

Flight-period • One brood. Scandinavia: mid-June to July. S Europe: June–August.

Habitat • Sheltered, alpine grassland, open moorland, & heaths. LHPs: alpine cinquefoil (*Potentilla thuringiaca*) & *Alchemilla glomerulans*.

Pyrgus cacaliae Dusky Grizzled Skipper

Distribution • C European Alps. Bulgaria (Rila Mts, Pirin Mts, & Stara Planina). Romania (Bucegi Mts). 1800–2800 m. Local.

Description • Male ups gc dark, greyish brown; upf markings small, often absent on uph. Female ups gc more brown. Both sexes: uns markings indistinct, giving overall 'washed-out' appearance – distinctive & characteristic.

Flight-period • June–August in one brood.

Habitat • Sheltered sites in open, alpine grassland, often near low bushes or scrub. LHPs: cinquefoils (*Potentilla*) & sibbaldia (*Sibbaldia procumbens*).

Pyrgus centaureae
Northern Grizzled Skipper

Distribution • N Fennoscandia: 60°N to North Cape. 0–950 m. Widespread & local, generally absent from coastal districts.

Description • Ups gc dark grey, often with whitish scaling; white markings well-developed; unh white veins prominent & distinctive.

Flight-period • Mid-June to July in one brood. Emergence date depends on altitude, latitude, & weather conditions.

Habitat • Marshes, bogs, & damp heath containing an abundance of LHP, cloudberry (*Rubus chamaemorus*).

Spialia sertorius **Red-underwing Skipper**

Distribution • Portugal, through S Belgium & NW Germany, Corsica & Sardinia to W Czech Republic, Austria, Italy, Slovenia, & N Croatia. Widespread & common. 0–1650 m.

Description & variation • First brood: unh gc creamy, light olive. Second brood: unh reddish – coloration distinctive & characteristic. Resembles *S. orbifer* closely, but distribution of two species not known to overlap. § Corsica & Sardinia (0–1500 m), *S. s. therapne*. Resembles nominate form, but smaller; upf cell-mark roughly square; unh marginal markings reduced.

♂ S Spain (2nd brood)

Flight-period • April–June & July to early September in two broods.

Habitat • Grassy, flowery places, usually amongst low bushes, rocks, or in open woodland. LHPs: principally salad burnet (*Sanguisorba minor*).

Spialia orbifer
Orbed Red-underwing Skipper

Distribution • Sicily. E Czech Republic through S Poland to European Turkey & Greece, including Ionian islands & most E Aegean islands. 0–2000 m. Widespread & common.

Description • First brood: unh gc creamy, light olive. Second brood: unh reddish or rust-brown – coloration distinctive & characteristic. Resembles *S. sertorius* closely but distribution of two species not known to overlap.

♂ NE Greece (1st brood)

Flight-period • Mid-April to June & mid-July to August in two broods.

Habitat • Grassy, flowery places, usually amongst low bushes, rocks, or in open woodland. LHPs: principally salad burnet (*Sanguisorba minor*).

♀ N Greece (2nd brood)

Spialia phlomidis **Persian Skipper**

Distribution • S Croatia to Bulgaria & S Greece.
650–1650 m, generally below 1000 m. Absent from
Mediterranean islands. Sporadic & usually very uncommon.

Description • Male ups gc black; white markings well
defined; base of costa pale buff; uph discoidal white spot large;
unh gc pale olive; discal white markings forming a distinctive
band.

Flight-period • Late May to June in one brood.

Habitat • Hot, dry, often rocky places; dry
grassland with sparse, low-growing scrub.
LHP: possibly *Convolvulus*.

Behaviour • Males frequently visit damp
ground to feed.

♂ NW Greece

Muschampia tessellum **Tessellated Skipper**

Distribution • S Balkans, NW Greece, & SE Aegean island
of Simi (0–100 m). Generally 800–1100 m. Very sporadic but
often locally very common.

Description • Male ups gc grey; pattern of white markings
distinctive & well-developed; unh gc yellowish, sometimes
pale, delicate green; white markings well-defined &
characteristic. Female similar; ups gc dark brown.

Flight-period • Mid-May to mid-August in one prolonged
brood.

Habitat & Behaviour • Open, grassy,
places with an abundance of flowers,
especially tufted vetch (*Vicia cracca*),
Achillea, & thyme (*Thymus*), upon which
both sexes are fond of resting & feeding.
LHP: *Phlomis samia*.

♂ NW Greece

♂ NW Greece

Muschampia cribrellum **Spinose Skipper**

Distribution • E Hungary. Romania. Republic of
Macedonia. 800–850 m. Extremely sporadic, very local & very
uncommon.

Description • Resembles *M. tessellum*, but smaller. Ups &
uns white markings more conspicuous; upf with two pairs of
elongate white spots in s1b – most reliable distinguishing
feature.

Flight-period • Mid-May to mid-June in one brood.

Habitat • Dry, flowery grassland with scattered bushes.
LHP(s): uncertain.

Muschampia proto **Sage Skipper**

Distribution • Portugal & Spain (widespread & common) through S France, S Italy, N Sicily, to S Balkans & Greece, including Kithira, Karpathos, & Simi. 0–1600 m. Very sporadic & local in central southern range.

Description & variation • Ups gc olive grey, often with paler scaling. Uns gc seasonally variable, especially in female: greenish in spring emergence, brownish, orange or pinkish mid- to late summer.

♀ NW Greece

Flight-period • April–October in one prolonged emergence.

Habitat • Hot, dry, flowery places, often amongst mixed scrub, often dominated by LHP. LHPs: Jerusalem sage (*Phlomis fruticosa*), *P. lychnitis*, & *P. herba-venti*.

♀ NW Greece

Carcharodus alceae Mallow Skipper

Distribution • S Europe, including most Mediterranean islands. 0–2000 m. Widespread & common. Appears absent in coastal districts of SW Iberian peninsula from Lisbon to Cadiz: apparently replaced entirely by *C. tripolinus*.

Description • Indistinguishable from *C. tripolinus* without reference to male genitalia, but distribution of two species not known to overlap.

Flight-period • Early April to October in 3+ broods.

Habitat • Open flowery places, often with long grasses & light scrub; also in hot, dry, rocky places with sparse vegetation. LHP: principally common mallow (*Malva sylvestris*).

♂ NW Greece   ♀ NW Greece

Carcharodus tripolinus
False Mallow Skipper

Distribution • S Portugal & S Spain: confirmed in coastal districts from Lisbon (Estoril) to Cadiz. Exact distributional relationship with *C. alceae* unknown.

Description • Indistinguishable from *C. alceae* without reference to male genitalia, but distribution of two species not known to overlap.

Flight-period • March–September in 2+ broods.

Habitat • Hot, dry, flowery, grassy places, rocky gullies, & slopes. LHP: common mallow (*Malva sylvestris*).

Conservation • Intense human activity on south coast of Iberian peninsula poses serious threat.

Carcharodus lavatherae **Marbled Skipper**

Distribution • From Spain through S Europe to European Turkey & N Greece. 200–1600 m. Sporadic & local.

Description & variation • Ups greenish or yellowish brown with darker marbling; upf inner margin often conspicuously reddish brown or dull orange; unh gc whitish or pale, creamy yellow; markings indistinct. § S Balkans & Greece, *C. l. tauricus*. Ups gc greyish or greyish brown; uns gc chalky white.

♂ NW Greece

Flight-period • Mid-May to late July in one brood. Emergence date depends on locality.

Habitat • Hot, rocky gullies or dry grassy banks with sparse scrub or scattered trees. LHPs: yellow woundwort (*Stachys recta*), downy woundwort (*S. germanica*), field woundwort (*S. arvensis*), & *S. plumosa*.

Behaviour • In very hot conditions, males may gather in large numbers to take moisture from damp soil.

♀ N Greece

♀ colour variant N Greece

Carcharodus boeticus
Southern Marbled Skipper

Distribution • N Portugal, Spain (south of Cantabrian Mts), through SE Pyrenees, S France, SW Switzerland, NW Italy, Apennines, & Sicily. Widespread & common in western range, progressively more sporadic & local in eastern range. 500–1600 m.

Description & variation • Unh white markings arranged in a distinctive, reticulate pattern (cf. *C. lavatherae*). Ups gc becomes progressively paler in later broods, culminating in a light sandy brown in late summer.

♂ E Spain

Flight-period • Switzerland: July in one brood. S Spain: May, June–July, & August–September in 2–3 broods, according to locality.

Habitat • Hot, dry, rocky places with sparse vegetation & scrub. LHP: possibly white horehound (*Marrubium vulgar*e).

Carcharodus stauderi
Eastern Marbled Skipper

Distribution • Greece: in Europe, known only from E Aegean islands of Kos, Simi, & Rhodes. Near sea-level. Apparently, very local.

Description • Resembles *C. boeticus* very closely, but not known to occur within its range.

Flight-period • Data limited. Records relate to May–June, probably in 2+ broods.

Habitat • Hot, dry, rocky, flowery places. LHPs: white horehound (*Marrubium vulgare*) & *Phlomis cretica*.

Carcharodus flocciferus
Tufted Marbled Skipper

Distribution • Spain, through S France, Italy, Sicily, & S Poland, to C & SE Balkans & N Greece. 1000–2000 m. Sporadic, very local, & generally uncommon.

Description • Resembles *C. orientalis*, but usually larger. Ups gc darker grey; unh gc darker, often with bluish or violet tones in fresh specimens; whitish stripe & discal spot prominent & distinctive. *C. flocciferus* & *C. orientalis* very rarely occupy same habitat.

Flight-period • W & C Europe: late May to June & late July to August in two broods. SE Europe: early July to mid-August in one brood.

♂ N Greece

Habitat • Grassy, flowery slopes or meadows, rocky gullies, sometimes in woodland clearings, often in damp places. LHPs: yellow hedge woundwort (*Stachys recta*), alpine woundwort (*S. alpina*), betony (*S. officinalis*), downy woundwort (*S. germanica*), marsh woundwort (*S. palustris*), hedge woundwort (*S. sylvatica*), & *S. scardica*.

Behaviour • Both sexes fond of resting on tops of tall flower stems, especially those of LHPs (cf. *C. orientalis*).

♀ N Greece

Carcharodus orientalis
Oriental Marbled Skipper

Distribution • Hungary, W & S Balkans, European Turkey, & Greece, including Corfu, Kithira, Andros, Egina, Evia, Kea, Skyros, Limnos, Lesbos, Samos, Kos, & Kalimnos. 25–1650 m. Sporadic, especially in northern range, locally common in southern range.

Description & variation • Uns gc pale with obscure white markings (cf. *C. flocciferus*). Size & ups coloration variable.

Flight-period • April–August in 2–3 broods.

Habitat • Hot, dry, rocky, flowery places, often amongst sparse scrub or bushes. LHP (S Greece): *Stachys* species.

Behaviour • Males often sit on stones or soil (cf. *C. flocciferus*).

♂ S Greece

Erynnis tages **Dingy Skipper**

Distribution • S Europe to Ireland, NE Scotland, S Scandinavia, & Lithuania. Absent from Mediterranean islands except Corfu. 50–2000 m. Generally widespread & common, but very rare & local in Holland, N Belgium, & Lithuania. In decline in Ireland & Britain.

Description & variation • Ups gc variable, pale brown to dark, greyish brown with dark, greyish-blue overtones; markings variable, obscure to prominent. § W Ireland, f. *baynesi*. Ups light brown; markings prominent. Darker forms appear to associate with cooler conditions at higher altitudes.

♂ NW Greece

Flight-period • N & C Europe: late April to mid-June, usually in one brood. S Europe: early April to early June & late June to late August; at least partially double-brooded. In hot summers, partial second brood may occur in N Europe, including S England.

Habitat • Damp or dry grassy, flowery places. LHPs: principally bird's-foot trefoil (*Lotus corniculatus*).

Behaviour • In hot conditions, males sometimes visit damp ground to drink.

♀ SE France

Erynnis marloyi **Inky Skipper**

Distribution • Albania. Republic of Macedonia. European Turkey. Greece, including Corfu, Lesbos, Chios, & Samos. 600–2000 m, generally below 1500 m. Very sporadic, locally common.

Description • Ups & uns gc very dark brown, almost black in some specimens. Unlikely to be confused with any other species.

♂ S Greece

Flight-period • Mid-May to late June in one brood.

Habitat • Hot, dry, rocky places, invariably on limestone or other calcareous rocks.

Behaviour • Females fond of feeding on flowers of thyme (*Thymus*). Especially wary & readily disturbed. Males often assemble, sometimes in large numbers, at summits of small hills.

♀ S Greece

Heteropterus morpheus
Large Chequered Skipper

Distribution • N Spain, to E Denmark (Lolland & Falster), SE Sweden (Skåne), & E Latvia, through Italy & C Balkans to SE Bulgaria & European Turkey. 0–1000 m. Channel Islands (Jersey). Extremely sporadic & local, especially in western range.

♂ N France

Description • Ups dark brown; upf costa with 3–4 small, yellow spots, better developed in female.

Flight-period • Late June to July in one brood.

Habitat • Damp or marshy places with tall grasses, usually associated with woodland. LHPs: grasses, including slender false brome (*Brachypodium sylvaticum*) & purple moor-grass (*Molinia caerulea*).

Conservation • As with most wetland species, habitats at risk from drainage.

Carterocephalus palaemon
Chequered Skipper

Distribution • Pyrenees to Arctic Circle & N Greece (Rhodopi Mts). Locally common in W Scotland. Extinct in England. 200–1600 m. Locally common; very sporadic near limits of range.

♂ N Greece

Description • Ups gc dark brown; pattern of yellowish-gold markings very distinctive.

Flight-period • Mid-May to July in one brood.

Habitat • Dry or damp grassy, woodland clearings. LHPs: grasses, including purple moor-grass (*Molinia caerulea*), slender false brome (*Brachypodium sylvaticum*), tor grass (*B. pinnatum*), & hairy brome (*Bromos ramosus*).

Behaviour • Both sexes strongly attracted to nectar of bugle (*Ajuga repans*, *A. pyramidalis*, & *A. genevensis*) &, in Scotland, bluebell (*Endymion non-scriptus*). Often roosts on tall grass stems.

♂ N Greece

♀ S France

Carterocephalus silvicolus
Northern Chequered Skipper

Distribution • NE Germany, through N Poland to Baltic countries & N Norway. In Denmark, restricted to Lolland. Widespread in eastern range, often locally common. 0–200 m.

Description • Superficially similar to *C. palaemon*. Male ups yellow markings much more extensive & with a prominent additional yellow spot near costa of uph. Female ups similar, but yellow markings greatly reduced.

Flight-period • Late May to late June. Emergence date depends on locality & weather conditions.

Habitat • Damp, sunny, sheltered, flowery, woodland clearings. LHPs: grasses, including hairy brome (*Bromus ramosus*), slender false brome (*Brachypodium sylvaticum*), & crested dog's-tail (*Cynosurus cristatus*).

Thymelicus acteon **Lulworth Skipper**

Distribution • Most of Europe, including S England (restricted to Dorset coast). Recorded from Canary Islands (100–1000 m) Sicily, Elba, Corfu, & several Aegean islands, including Santorini, Kithira, Spetses, Crete, & Rhodes. 0–1600 m. Widespread & common.

Description & variation • Male ups gc dusky orangey brown; upf with yellowish streak in cell & often obscure pd yellowish spots in s3–9; thin, black sex-brand conspicuous. Female upf without sex-brand, pd yellowish spots usually better developed. Both sexes: uns gc yellowy orange. § Hottest regions of Mediterranean. Ups darker, tending to greenish or greyish brown. § Canary Islands, *T. a. christi*. Ups more brown; upf yellowy-orange markings well-defined.

♂ NE Greece

Flight-period • Most of Europe: mid-May to early August in one prolonged brood. Canary Islands: February to early October, reputedly in 2–3 broods.

Habitat • Grassy, flowery, often very hot places, usually amongst scrub. LHPs: grasses, including tor grass (*Brachypodium pinnatum*), slender false brome (*B. sylvaticum*), & small wood reed (*Calamagrostis epigejos*).

Thymelicus lineola **Essex Skipper**

Distribution • Most of Europe, including S England, Sicily, & Corsica. 0–2200 m. Widespread, locally very common.

Description • Resembles *T. sylvestris*, but generally smaller. Ups narrow, black marginal borders more prominent; black veins often conspicuous. Male upf sex-brand shorter. Uns of antennal tip black or dark brown.

Flight-period • May–August in one prolonged brood.

Habitat • Flowery places containing tall grasses. *T. sylvestris* almost always shares same habitats. LHPs: grasses, including

♀ N Greece

common cat's-tail (*Phleum pratense*), creeping soft-grass (*Holcus mollis*), & slender false brome (*Brachypodium sylvaticum*).

In recent decades, extension of range in S England appears to have been facilitated by grassy verges of new & highly integrated road systems.

Thymelicus sylvestris Small Skipper

Distribution • Most of Europe, including Wales, England, Sicily, Corfu, & several Aegean islands except Rhodes & Crete. 0–1900 m. Widespread, locally very common.

Description • Resembles *T. lineola*, but generally larger. Ups black veins usually less prominent. Upf male sex-brand longer. Uns of antennal tip orangey brown.

Flight-period • Late April to July in one brood. Emergence date depends on local/regional climate.

Habitat • Flowery places containing tall grasses. LHPs: grasses, including Yorkshire fog (*Holcus lanatus*), creeping soft-grass (*H. mollis*), common cat's-tail (*Phleum pratense*), & slender false brome (*Brachypodium sylvaticum*).

Behaviour • Females are often very fastidious in choice of egg-laying sites.

♂ NE Greece   ♀ NE Greece

Thymelicus hyrax Levantine Skipper

Distribution • Greece (Mt Parnassos & Askion Mts. 600–800 m) including Aegean islands of Samos, Chios, Lesbos, & Rhodes. 0–250 m. Apparently very sporadic & local, but possibly overlooked owing to confusion with allied species.

Description • Resembles *T. sylvestris* closely, but uns hw more greenish.

Flight-period • Late April to late June in one brood.

Habitat • Hot, dry, rocky places.

Behaviour • On Mt Parnassos, both sexes are greatly attracted to flowers of a small, bushy thyme (*Thymus*).

Hesperia comma **Silver-spotted Skipper**

Distribution • Most of Europe, including N Sicily, S England, & Lappland. 0–2300 m. Widespread, locally common.

Description & variation • Ups gc greyish brown with extensive discal & pd orange markings; upf with yellowish subapical & pd spots; unh whitish-silver spots well-defined, conspicuous. Male upf black sex-brand well-defined. § Colder regions (Lappland & high altitudes in C Europe). Smaller; ups darker, tending to brown; uns gc darker green.

♀ NW Greece

Flight-period • Late June to mid-September in one brood.

Habitat • Open, flowery places with short grass containing an abundance of LHP. LHPs: principally sheep's fescue (*Festuca ovina*).

♀ NW Greece

Ochlodes venatus **Large Skipper**

Distribution • Most of Europe, including Sicily, Corfu, Wales, England, & S Scotland. Generally 0–1800 m; above 1000 m in S Spain. Widespread & common.

Description & variation • Ups gc dusky brown with extensive discal & pd orange markings; unh gc greenish with yellow discal spot & pd spots; bright orange area in anal angle extending almost to wing-base. Male upf black sex-brand prominent. § At high altitudes & in colder, northern localities. Smaller; ups & uns slightly darker.

♂ N Greece

Flight-period • Generally June–August in one brood.
Spain: recorded in mid-May, June, July, & late August; possibly in two broods.

Habitat • Grassy, bushy, woodland margins & clearings. Most habitats are humid & often contain an abundance of ferns, especially bracken (_Pteris aquilina_), & blackberry (_Rubus fruticosus_). LHPs: grasses, including cock's-foot (_Dactylis glomerata_), slender false brome (_Brachypodium sylvaticum_), & common cat's-tail (_Phleum pratense_).

Behaviour • Both sexes strongly attracted to bramble blossom, & often rest or bask on its leaves.

♂ NW Greece

Gegenes nostrodamus
Mediterranean Skipper

Distribution • Very local & sporadic, mostly in Mediterranean coastal regions from S Portugal & Spain to S Balkans & Greece, including Mediterranean islands of Mallorca, Corsica, Sardinia, Elba, Sicily, Spetses, & Crete. 0–250 m. Greece: NW Pindos Mts. 400–1200 m. Generally does not occur in same habitat as *G. pumilio*, but they fly together in some localities on Crete.

Description • Male ups brown, unmarked (cf. *Borbo borbonica* & *Pelopidas thrax*); unf pale, greyish, olive brown, with 2–3 indistinct spots; unh pale sandy, olive brown, shading to white towards inner margin; unh cilia-like hairs on costa, long & dense (cf. *G. pumilio*). Female ups, greyish, olive brown; upf with distinct pd spots; unf with pale discal spots; unh, unmarked (cf. *G. pumilio*, *B. borbonica*, & *P. thrax*).

Flight-period • Late April to October in usually two, sometimes three, broods, depending on locality. Usually very scarce in first brood.

Habitat • Hot, dry, stony gullies, often in flood-plains of rivers, usually amongst sparse vegetation. LHPs: grasses.

Behaviour • Flight very fast & low. Males often sit on stones or soil in full sun. Extremely wary but returns quickly to original resting site when disturbed. Females perch on tall grass or flower stems in early morning; much less in evidence at other times.

Conservation • As with all species in coastal districts of Mediterranean in close proximity to areas of intensive, human activity, habitats are especially vulnerable.

♂ S W Spain

Gegenes pumilio **Pigmy Skipper**

Distribution • Very local in mostly Mediterranean coastal regions from S Spain to Greece (0–400 m), including Mediterranean islands of Mallorca, Corsica, Sardinia, Elba, Sicily, Malta, Corfu, & several Aegean islands, including Crete, Spetses, Samos, Kos, & Rhodes; also inland regions of NW Greece (500–1800 m).

♂ N Greece

Description • Ups very dark brown; uhf & unh, pale brown with white, poorly defined pd spots. Male ups without markings (cf. *Borbo borbonica* & *Pelopidas thrax*); unh cilia-like hairs on costa, short & sparse (cf. *G. nostrodamus*). Female upf with pale, poorly defined pd spots.

Flight-period • April to late October in 2+ broods. Generally very scarce in first brood.

Habitat, behaviour, & conservation • As for *G. nostrodamus*.

♂ S Greece

Borbo borbonica **Zeller's Skipper**

Distribution • SW Spain & Gibraltar. 0–50 m. Very local in coastal districts.

Description • Superficially similar to *Gegenes nostrodamus* & *G. pumilio*. Ups dark, greyish, olive brown; upf pd spots hyaline, spot in s1b yellow; unf olive brown; unh yellowish brown, with three, small, well-defined, pale pd spots; palpi buff. Female larger.

Flight-period • Records span June–November, probably in 2+ broods. Most observations relate to August–October.

Habitat • Hot, dry, rocky, coastal gullies. LHPs: grasses.

Behaviour • Reputedly an occasional migrant.

Conservation • As for *G. nostrodamus*.

Pelopidas thrax **Millet Skipper**

Distribution • Greece (Samos & Rhodes). 0–75 m. Possibly overlooked on other eastern Aegean islands. Very local, not common.

Description • Ups dark, greyish, olive brown; upf hyaline spots distinctive. Male upf narrow, white sex-brand distinctive (cf. *Gegenes nostrodamus* & *G. pumilio*); unh without pd spots (cf. *G. pumilio*). Female upf spots better developed; uph & unh with 3–4 pd white spots (cf. female *G. nostrodamus* & *G. pumilio*).

Flight-period • Records for Greece relate to June. Turkey: May–July & late September to mid-October in two broods.

Habitat • Hot, dry, grassy places in low-lying, coastal districts. LHPs: grasses.

Behaviour • Flight low & powerful. Reportedly a migrant.

Glossary

Abdomen The part of the body behind the thorax.

Aestivation A state of torpor (dispause) in summer heat or drought.

Anal angle Small apical area enclosed by inner and outer margins of the hind-wing.

Androconia (sing. **androconium**) Specialized wing-scales (often called scent-scales) in male butterflies, possessing gland cells containing special chemicals for attracting females.

Antennae (sing. **antenna**) Paired, jointed sensory organs – clubbed in the case of butterflies – arising from the head of an insect.

Anterior Towards the head in reference to axis of head, thorax, and abdomen. Applicable to adult, larvae, or pupa. (Cf. posterior.)

Apex The point of coincidence of the costal and outer margins.

Apical area Of fore-wing, area just inside and contiguous with apex. (Cf. subapex.)

Basad Towards the wing-base.

Basal Of the wing-base.

Caterpillar See larva.

Calcareous Referring to rocks/soils having an alkaline (basic) reaction.

Cell In reference to wings, the generally closed area defined by the subcostal, medial, and discoidal veins in the basal and discal areas of the fore-wings and hind-wings: due to the absence of one or more elements of the discoidal vein, the cell on the hind-wing is sometimes open. (See also discocellular.)

Chrysalis See pupa.

Club The thickened, terminal part of the antenna.

Colony A small, locally isolated population.

Costa The front (leading) edge of fore-wing or hind-wing.

Discal The central portion of wing from costa to inner margin: hence, discal band, discal markings, etc.

Discoidal Of the area associated with the cell: hence discoidal spot.

Distal The point furthest away from the centre of the body.

Dorsal Of the back – upper surface of the body (Cf. ventral.)

Ecology The study of the relationships of animals and plants with each other and their environment.

Elongate Of extended or lengthened form – stretched.

Endemic Restricted to a well-defined region – found nowhere else.

Family A basic unit of taxonomic classification, usually comprising an assemblage of genera considered to be closely related on account of certain shared characters.

Fennoscandia Geographical region of North-west Europe comprising most of Scandanavia and Finland and parts of the Soviet Union west of the White Sea.

Form Any taxonomic unit subordinate to subspecific classification applicable to ecological, seasonal, or sexually dimorphic/polymorphic forms.

Frons The area between the eyes on the front the head ('face'), often bearing a hair-like tuft.

Fuscous Smokey, greyish brown.

Genitalia The male or female sexual organs, located in the terminal abdominal segments, by which means the spermatophore of the male is transferred to the female during copulation.

Genus (pl. **genera**) A basic unit of taxonomic classification, usually comprising a number species considered to be more closely related to each other than to other species of other genera.

Hair-collar Ring of hair-like scales between head and thorax.

Hibernation The dormant stage in which an animal passes the winter. (See also diapause.)

Honey-gland A gland located in the dorsal region of the 7th segment, producing a sweet secretion attractive to ants. Present in the larvae of many species of Lycaenidae.

Hybrid The progeny of two species arising from cross-fertilization.

Inner margin Of the wings, the margin closest to the body.

Internal The point closest to the body.

Larva (pl. **larvae**) The second (growth) stage in the development of an insect.

Lunule A crescent-shaped mark.

Macular Spotted.

Marginal Of fore-wing or hind-wing, the wing area contiguous with the outer margin.

Mediobasal Central transverse line of basal area.

Mediodiscal Central transverse line of wing – bisecting the discal area.

Migrant Of butterflies, survival strategy depending partly or wholly on dispersion and establishment of temporary breeding colonies over an extensive geographical area. (See Dispersion.)

Mimic (noun or verb) A species (animal or plant) bearing a close resemblance, in part or whole, to another, used as a model; to adopt, in part or whole, the superficial characters of another species.

Morphology The study of structure and form.

Nominate form Having characters corresponding to the type specimen upon which species description is based – the typical form, often referred to as the nominotypical form.

Ocellus (pl. **ocelli**) An 'eye-spot': a rounded spot (often black), usually with a central white pupil or pale spot, often enclosed by a pale or coloured annulus (ring). When the pupil is absent, the spot is said to be 'blind'.

Palp (pl. **palpi**) One of a pair of sensory organs located on the front of the head in an adult butterfly.

Photoperiod Day-length, period between dawn and dusk.

Population Individuals of a species living together or in sufficiently close proximity to sustain, over time, a high probability of maintaining uniform (within limits of normal variation) genetic character.

Postdiscal Of the wings, the area between the discal and submarginal areas.

Posterior Towards the abdominal extremity relative to axis of head, thorax, and abdomen.

Proboscis The feeding tube of the adult butterfly.

Race A distinctive population of a species approximating, but generally considered subordinate to, a subspecies.

Range The total (unless otherwise indicated) geographic area of species or subspecies: distribution need not be continuous.

Scandinavia Geographical region of Europe comprising Denmark, Norway, and Sweden.

Sex-brand A grouping of androconia – often in conspicuous patches.

Space Of the wings, an area of wing-membrane between two veins.

Species (pl. **species**) A basic unit of formal taxonomic classification referring to a group of individuals of an organism capable of interbreeding and producing healthy, fertile offspring. Such groups are incapable of cross-breeding with other groups to produce healthy, fully fertile offspring.

Sphragis A hard structure formed on the ventral surface of the terminal abdominal segments of a fertilized (female) butterfly to prevent further copulation.

Stria (pl. **striae**) A narrow line or streak.

Subapex Of the fore-wing, the area inside the apex: hence subapical area.

Submarginal Of fore-wing or hind-wing, the wing area between that just inside the outer margin and the postdiscal area. (Closest to the outer margin is the marginal area.)

Subspecies A population occupying a distinct geographical region, separate from other populations of the same species, and having constant and clearly different characters.

Symbiosis Living together: a close and often obligatory association of two species: e.g. the associations of ants and lycaenid larvae.

Thorax The middle section of an insect's body: in the adult butterfly, the clearly separate portion bearing the wings and legs.

Tundra Grassy, treeless, Polar regions with permanently frozen subsoil (permafrost). In lower latitudes, tundra-like zones, conforming closely to Arctic climatic conditions, occur at high altitudes.

Variety A poorly defined subunit of classification below the rank of subspecies.

Vein In an insect's wing, the semi-rigid tubes supporting the wing membrane.

Venation The arrangement of veins in an insect's wing.

Ventral Below the central plane of the body: hence ventral surface of the wings. (Cf. Dorsal.)

Bibliography

Abadjiev, S. 1992. *Butterflies of Bulgaria. Part 1. Papilionidae and Pieridae.* Veren Publishers, Sofia.

Abadjiev, S. 1993. *Butterflies of Bulgaria. Part 2. Nymphalidae: Libytheinae and Satyrinae.* S. Abadjiev, Sofia.

Abadjiev, S. 1995. *Butterflies of Bulgaria. Part 3. Nymphalidae: Apaturinae and Nymphalinae.* S. Abadjiev, Sofia.

Bink, F. A. 1992. *Ecologische Atlas van de Dagvlinders van Noordwest-Europa.*

Chiavetta, M. 2000. *Le Farfalle d'Italia-Atlante Biogeografico.* Editoriale Grasso, Bologna. (In Italian and English)

Chinery, M. 1998. *Butterflies of Britain and Europe.* HarperCollins, London.

Ebert, G. (ed.) 1993. *Die Schmetterlinge Baden-Württembergs.* Vols 1 and 2. Eugen Ulmer, Stuttgart.

Ford, E. B. 1957. *Butterflies.* Collins, London.

Geiger, W. (ed.) 1987. *Les papillons de jour et leurs biotopes.* Ligue Suisse pour la Protection de la Nature, Bâle. (Also in German.)

Gómez-Bustillo, M. R. and Fernández-Rubio, F. 1974. *Mariposas de la Peninsula Ibérica: Ropalóceros.* Vols 1 and 2. Ministerio de Agricultur, Madrid.

Heath, J. (ed.) 1970. *Provisional atlas of the insects of the British Isles. 1. Lepidoptera Rhopalocera: Butterflies.* Biological Records Centre, Huntingdon.

Heath, J. (ed.) 1976. *The moths and butterflies of Great Britain and Ireland.* Vol. 1. Blackwell Scientific and Curwen, Oxford and London.

Heath, J. 1981. Threatened Rhopalocera (butterflies) in Europe. *Nature Environ. Ser.,* **23,** 1–157.

Henriksen, H. J. and Kreutzer, I. 1982. *The butterflies of Scandinavia in nature.* Skandinavisk Bogforlag, Odense.

Higgins, L. G. 1975. *The classification of European butterflies.* Collins, London.

Higgins, L. G. and Riley, N. D. 1984. *A field guide to the butterflies of Britain and Europe.* Collins, London.

Howarth, T. G. 1973. *South's British butterflies.* Warne, London.

Lafranchis, T. 2000. *Les Papillons de jour de France, Belgique et Luxembourg et leurs chenilles.* Collection Parthénope, Mèze.

Manley, W. B. L. and Allcard, H. G. 1970. *A field guide to the butterflies and burnets of Spain.* Classey, Hampton.

Pamperis, L. N. 1997. *The butterflies of Greece.* Bastas-Plessas, Athens.

Polunin, O. 1969. *Flowers of Europe.* Oxford University Press, London, New York, and Toronto.

Polunin, O. 1980. *Flowers of Greece and the Balkans.* Oxford University Press, Oxford, New York, Toronto, and Melbourne.

Schaider, P. and Jak ic, P. 1988. *Die Tagfalter von jugoslawisch Mazedonien.* Schaider, Munich.

Strid, A. 1986. *Mountain flora of Greece.* Vol. 1. Cambridge University Press, Cambridge, London, New York, New Rochelle, Melbourne, Sydney.

Thomas, J. A. and Lewington, R. 1991. *The butterflies of Britain and Ireland.* Dorling Kindersley, London.

Thomson, G. 1980. *The butterflies of Scotland.* Croom Helm, London.

Times Books. 1994. *The Times atlas of the world – comprehensive edition.* HarperCollins, London.

Tolman, T. W. 1997. *A field guide to the butterflies of Britain and Europe.* HarperCollins, London. (Also in French, German, Danish, and Dutch.)

Tutin, T. G., *et al.* (eds). 1964–80. *Flora Europaea.* Vols 1–5. Cambridge University Press, Cambridge.

Index

Index